Agricultural Production, Marketing and their Management

The Author

Dr. Kiros Ventura, (Ph.D. Agricultural Economics) has long experience in teaching and research at the college of Business and Economics, Hawassa University, Awassa, Ethiopia. His research areas include Agricultural production, Economics, Farming System Approach, marketing, Management and International Trade of Agriculture.

Agricultural Production, Marketing and their Management

Dr. Kiros Ventura

2026

Agri Horti Press

New Delhi-110 084

ISBN 978-93-87642-52-2 (Hardbound)

Published By

Agri Horti Press

New Delhi

E-mail: agrihortipress@gmail.com

Digitally Printed at : Replika Press Pvt. Ltd.

PREFACE

The study of agricultural production, marketing and their management comprises all the operations, and the agencies conducting them, involved in the movement of farm-produced foods, raw materials and their derivatives, for example textiles, from the farms to the final consumers, and the effects of such operations on farmers, middlemen and consumers. Agricultural marketing is the study of all the activities, agencies and policies involved in the procurement of farm inputs by the farmers and the movement of agricultural products from the farms to the consumers. The agricultural marketing system is a link between the farm and the non-farm sectors. It involves all the aspects of market structure or system, both functional and institutional, based on technical and economic considerations, and includes pre and post-harvest operations, assembling, grading, storage, transportation and distribution. A dynamic and growing, agricultural sector requires fertilizers, pesticides, farm equipments, machinery, diesel, electricity and repair services which are produced and supplied by the industry and non-farm enterprises. The expansion in the size of farm output stimulates forward linkages by providing surpluses or food and natural fibers which require transportation, storage, milling or processing, packaging and retailing to the consumers. Agricultural marketing plays an important role not only in stimulating production and consumption, but in accelerating the pace of economic development. The agricultural marketing system plays a dual role in economic development in countries whose resources are primarily agriculture. Increasing demands for money with which to purchase other goods leads to increasing sensitivity to relative prices on the part of the producers, and specialization in the cultivation of those crops on which the returns are the greatest, subject to socio-cultural, ecological and economic constraints. This book is designed as an aid for students and professionals of agricultural marketing, agricultural economics, agribusiness, management and business studies. Further queries, constructive suggestions and criticism for improvement of the book are always welcome and shall be thankfully acknowledged.

Agricultural production, marketing and their management plays an important role not only in stimulating production and consumption, but in accelerating the pace of economic development. The agricultural marketing system pays a dual role in economic development in countries whose resources are primarily agriculture. This book presents information on both theoretical and practical issues on farm products and marketing management, agricultural production and marketing, agricultural finance and marketing, fibre crops, agricultural marketing in developing countries, agricultural marketing system and promotional management in agriculture. This is available source book for the students and professionals of agricultural marketing, agricultural economics, agribusiness management and business studies.

Kiros Ventura

Contents

1

Introduction

Organised marketing of agricultural commodities has been promoted in the country through a network of regulated markets. Most of the State governments and Union Territories have enacted legislations (APMC Act) to provide for regulation of agricultural produce markets. While by the end of 1950, there were 286 regulated markets in the country, today the number stands at 7,521 (31.3.2005).

Besides, the country has 27,294 rural periodical markets, about 15 per cent of which function under the ambit of regulation. The advent of regulated markets has helped in mitigating the market handicaps of producers/ sellers at the wholesale assembling level. but, the rural periodic markets in general, and the tribal markets in particular, remained out of its developmental ambit.

Agriculture sector needs well functioning markets to drive growth, employment and economic prosperity in rural areas of the country. In order to provide dynamism and efficiency into the marketing system, large investments are required for the development of post harvest and cold chain infrastructure nearer to the farmers' field.

A major portion of this investment is expected from the private sector, for which an appropriate regulatory and policy environment is necessary. Alongside, enabling policies need to be put in place to encourage procurement of agricultural commodities directly from farmers' field and to establish effective linkage between the farm production and the retail chain and food processing industries.

Accordingly, amendment to the State APMC Act for deregulation of marketing system in the country is suggested to promote investment in marketing infrastructure, motivating corporate sector to undertake direct marketing and to facilitate a national integrated market.

Concept and Importance

The study of agricultural marketing comprises all the operations, and the agencies conducting them, involved in the movement of farm-produced foods, raw materials and their derivatives, such as textiles, from the farms to the final consumers, and the effects of such operations on farmers, middlemen and consumers. Agricultural marketing is the study of all the activities, agencies and policies involved in the procurement of farm inputs by the farmers and the movement of agricultural products from the farms to the consumers. The agricultural marketing system is a link between the farm and the non-farm sectors. It involves all the aspects of market structure or system, both functional and institutional, based on technical and economics considerations, and includes pre and post-harvest operations, assembling, grading, storage, transportation and distribution. A dynamic and growing, agricultural sector requires fertilizers, pesticides, farm equipments, machinery, diesel, electricity and repair services which are produced and supplied by the industry and non-farm enterprises. The expansion in the size of farm output stimulates forward linkages by providing surpluses or food and natural fibres which require transportation, storage, milling or processing, packaging and retailing to the consumers.

Importance

Agricultural marketing plays an important role not only in stimulating production and consumption, but in accelerating the pace of economic development. The agriculture marketing system plays a dual role in economic development in countries whose resources are primarily agricultural. Increasing demands for money with which to purchase other goods leads to increasing sensitivity to relative prices on the part of the producers, and specialisation in the cultivation of those crops on which the returns are the greatest, subject to socio-cultural, ecological and economic constraints.

It is the marketing system that transmits the crucial price signals:

- Agricultural Marketing is one of the manifold problems, which have direct bearing upon the prosperity of the cultivators, as India is an agricultural country and about 70 per cent of its population depends on agriculture.
- Most of the total cultivated area (about 76 per cent) is to under food grains and pulses. Approximately 33 per cent of the output of food grains, pulses and hearly all of the productions of cash crops like cotton, sugarcane, oilseeds etc. are marketed as they remain surplus after meeting the consumption needs of the farmers. Development of technology, quick means of communication and transportation has introduced specialisation in agriculture.
- Agriculture supplies raw materials to various industries and therefore, marketing of such commercial crops like cotton, sugarcane, oilseeds etc. assumes greater importance.
- With the introduction of green revolution agricultural production in general and food grains in particularly has substantially

increased. Agriculture once looked as a subsistance sector is slowly changing to a surplus and business proposition.

- The interaction among producers, market functionaries, consumers and government that determine the cost of marketing and sharing of this cost among the various participants.
- The producer, middleman and consumer look upon the marketing process from their own individual point of view. The producer is primarily concerned with selling his products.
- Any increase in the efficiency of the marketing process, which results in lower costs of distribution at lower prices to consumers, really brings about an increase in the national income.
- A reduction in the cost of marketing is a direct benefit to the society.
- Marketing process brings a new varieties, qualities and beneficial goods to consumers and therefore, marketing acts as a line between production and consumption.
- Scientific, systematic marketing stabilises the price level.
- An improved marketing system will stimulate the growth of number of agrobased industries mainly in the field of processing.
- A marketing system can become a direct source of new technical knowledge and induce farmers to adopt upto date scientific methods of cultivation.

Marketing is therefore, playing an important role in the economic development and stability of a country.

INFRASTRUCTURE of AGRICULTURAL MARKETING

The Rationale: Infrastructure consists of a combination of national assets which sustain the addition of place, time and form utilities to the products and services. These include apart from the Government institutions and organisations, roads, railways, warehouses, market yards, cold stores, processing units, research and training institutions, means of communication and transportation including air cargo, sea cargo etc. The basic rationale of any infrastructure is the sustenance it provides to production activity, income generation and social service supplies. It has also positive effect on income distribution because low per capita infrastructure limits the access of small and marginal farmers to the market.

The relationship between agricultural development and investment in infrastructure has been long recognised. Roads stimulate agricultural change and modernisation not only through their immediate effects on relative prices and marketing opportunities but also through backward linkages. The roads open up opportunities for commercial agriculture and encourage shifts to production of higher value, transport - sensitive products (fruits, vegetables, dairy, poultry and marine products). Roads also improve access of the people to extension agents, banks, markets and health services. Market infrastructure is important not only for the performance of various marketing functions and expansion of the size of the market but also for transfer of appropriate price signals leading to improved marketing efficiency. Infrastructure facilities lead to reduction in marketing costs which is crucial for increasing the realisation of growers and reducing the costs to the consumer.

The infrastructural facilities can be classified as physical and institutional. The roads, railways, transport facilities, electrification and storage structures are physical infrastructure whereas cooperatives, local self-government, banking institutions, extension agencies, marketing organisations and market intelligence net work are institutional infrastructure. For over four decades after independence, the public sector in India held a monopoly in the provision of most of the infrastructures. Till 1991, when the current period of economic reforms started, electricity, railways, roads, telecommunications, postal services and ports were among the sectors reserved for the public sector. However, after 1991, virtually all sectors of infrastructure have been opened to private investment. Nevertheless, for providing infrastructure in remote and difficult areas, the public sector would need to continue to play an important role.

Market Surplus

In order to assess the adequacy of agricultural marketing infrastructure in the country, it is imperative to estimate the availability of agricultural production and marketed surplus. Generally, there is positive association between production and marketed surplus. Several studies carried out by individual researchers and national and international organisations provide the projections of both demand and supply of agricultural commodities at different points of time. Projections for too distant a period involve several assumptions which may not hold good. The Committee has, therefore, used the projections of production of various farm products as given by Kumar and Mathur for the period 2006-07.

Projections of production and marketed surplus of various farm products for the year 2006-07, that even at the existing marketed surplus-output ratios, the quantities which the marketing system will be required to handle in future are quite large. For example rice output is projected at 103.5 million tonnes, meaning thereby that paddy output available for milling with rice milling sector would be around 155 million tonnes. The marketed surplus of all cereals taken together would be 102.74 million tonnes. As regards pulses, the marketed surplus is projected at 15.20 million tonnes which will require considerable increase in the pulse milling capacity. Reduction of proporation of population on agriculture would enable rise in market surpluses.

The marketed surplus of oilseeds is projected to go up from 17.19 million tonnes during 1999-2000 to 26.90 million tonnes during 2006-07. In the case of sugarcane, the marketed quantity is projected to reach 327.75 million tonnes. Raw cotton output is projected at 3.2 million tonnes during 2006-07. similarly fruit and vegetables marketing system will be required to handle 91.88 million tonnes of vegetables and 68.38 million tonnes of fruits.

As regards livestock products, it is projected that by 2006-07, the marketed surplus of milk would be 71.7 million tonnes, meat and eggs 6.0 million tonnes and of marine products 9.80 million tonnes. Considering all the perishables together (fruits, vegetables and livestock products), the marketed surplus is anticipated to go up by 43.4 per cent during the next seven years from 172.69 million tonnes during 1999-2000 to 247.76 million tonnes during 2006-07. The capacity to clean, grade, pack, process and transport these perishables would have to expand to handle the additional marketed quantities.

From the view point of complete supply chain, from farm to the market, the infrastructure for all types of perishable horticulture produce is required at following levels:

- Small pre-cooling units and or zero-energy cool chambers in the production areas where the field heat of the produce is to be removed at fast rate to bring down the temperature of the produce to the desired level before putting the product in the cold storages. The refrigerated transport units from the farm to the cold storages are also utilised as mobile pre-cooling units for this purpose;
- Collection Centers near to the farms;
- Medium to small cold storages having multi-product, multi-chamber facilities are the most popular segment where horticulture produce is stored as transit godowns;
- Specialised cold storage with facility of built in pre-cooling; high humidity and Controlled/Modified Atmosphere are required for storage of the produce for a longer period. These specialised storages are essential for extended shelf life of the produce and without these storages the requirement of storing the produce to meet the demand in the off season is not feasible;
- Other components like ripening chambers close to the markets and display cabinets at retail outlets;

- Linkages for conversion of fresh produce in other marketable forms;
- Integrated Pack Houses to serve farms in respective regions having an area of around 5000-10000 hectare. Farms associated with each of the centers would collect farm produce and bring them to common cold storage centers, where these products could be given treatments, such as washing, sorting, grading and packing. These products will then be preserved in the appropriate cold storage facility. The services of these centers will not only increase the value of the farm product, but will also remove most of the unwanted bio-degradable bio-mass from the horticulture products, which can be utilised as farm manure or even as cattle feed.

The electronic trading would be more appropriate form of direct marketing between different buyers and sellers. Every market committee should be provided with facility of electronic trading by setting up a special kiosk for the purpose. The young entrepreneur who can set up portals on their own to provide such facilities could be supported financially through a plan scheme. Their responsibilities would include to inform the buyers and sellers about online demand of different products; product specifications with regard to quality, pack size, packaging material, quantity and the time frame of supply; the transport cost involved and the marketing charges likely to be incurred in the market where the goods are to be delivered; facilities available to the farmer in the buying market; Re-handling of the produce, if necessary, in the supplying market to suit to the requirement of the buyer market; the rules and regulations of the destination market, if it is located outside the state at distant place, and other specific information as may be conducive for the seller to transact the business with the purchasers; and the legal provisions related to storage, transportation, phyto-sanitary requirements etc.

Telephone is the most convenient way of communication which can reduce cost of marketing besides meeting other needs in the villages. However, upto March 2001, only 59 per cent of 6,70,000 villages were connected by telephone. The tele-density *i.e.* number of telephones per hundred population was only 3.5 per cent compared to an average of 16 in the world and 60 in the developed world. Though, it has been targeted to increase this to 7 per cent by 2005 and 15 by 2010, the progress seems to be slow. There is a need to speed up the investment in this regard. Telephone connection technology by telephone lines seems to be slow and requires heavy investment for laying telephone lines. Therefore, the option of wireless technology (WCL), should be examined. The Committee recommends that all the remaining villages should be connected within five years. Also, at the same time e-communication should be encouraged either through village panchayat or private entrepreneurs. Such communication or Cyber Café or village kiosks can become information centers.

Rural Connectivity

Rural roads constitute one of the most important marketing infrastructure which reduce the cost of production and marketing by providing external economies to farmers, traders and public at large. It is well known that investment in infrastructure of this type has very high returns to the society. The status of rural roads in India indicates that only 47.83 per cent villages were covered with roads till mid 90s as can be seen from the Table below:

Table. Details of Road Connectivity in India

Group of villages with population	Villages connected with road	% age of villages connected	Villages not connected with roads	% age of villages not connected	Total
<1000	172062	37.45	280733	63.55	452795
1000-1500	44031	75.88	13904	24.12	57935
>1500	65698	91.73	5713	8.73	71411
Total	281791		300350		582142

Therefore, there is an urgent need for investment in providing connectivity to remaining villages. Considering average road length as 4 km., to connect each village with the main road, public yard or sub-yard, the total length of rural roads required to connect to remaining 3.7 lakh villages comes to around 14.8 lakh kms. If average cost is taken as ₹.5 lakhs per km, the investment requirement is ₹.74,000 crores. Though the Government of India has made special provision for link roads in rural areas under Prime Minister's programme there is a need for additional investment.

Physical Facilities in Markets

Agricultural produce markets established under market regulation programme have been playing an important role in providing market places to the farmers to dispose off their produce. These have also provided physical facilities and an institutional environment to the traders, processors and other market functionaries for conduct of their trading activities. The studies revealed that farmers, on an average, get a reasonably higher price by selling their produce in the regulated market yards compared to rural, village and unregulated wholesale markets. Most of the regulated market yards in the country at present lack facilities for handling the produce arriving there. The space for auction platform is less and the number of shops and godowns in the premises is small. It reduces the effective participation of traders. Absence of storage godowns at market level further perpetuates the problems of traders in general and continuous movement of goods in particular. The number of fruit and vegetables markets brought under regulation is small. Further the markets, which have been exclusively developed for handling of fruits and vegetable, do not have sufficient facilities for handling the produce available in the area.

The Directorate of Marketing and Inspection and several state governments have assessed the requirements of investment for development of market yards in respective states. Though several questions relating to the desirability of continuing with government sponsored market yards are being raised, the Expert Committee is of the view that creation of physical infrastructure at primary market places is absolutely essential irrespective of the institutional arrangements for managing these yards. The investment requirement for development of market yards/sub-yards during the next ten years is estimated at ₹.6026 crores.

Specialised Markets

Apart from general purpose markets, there is need for developing specialised markets for fruit and vegetables. It has been assessed that there are at least 241 such places in the country where fruit and vegetables markets should be developed. The infrastructure required for such markets depends on the volume of arrivals which in turn depends on the size of population to which these markets cater. The investment requirement for fruit and vegetables markets in the country is around ₹.970 crores. The details are tabled below:

Table. Investment Requirements for Development of Fruits and Vegetables markets (₹ In crores)

Size of Population	No. of Markets	Investment Required
1 to 2 lakhs	126	252
2 to 5 lakhs	72	288
Above 5 lakhs	43	430
Total	241	970

Farmers' Markets

Several State Governments have initiated a process of direct marketing by producers to the consumers. The states of Punjab, Haryana, Rajasthan, Tamil Nadu and Andhra Pradesh have established Apni Mandi in their areas. However, this has been promoted so far only at the state headquarters and at some district headquarters. There is a need to promote these in all the districts. The Committee recommends that farmers markets be promoted in all the districts of the country to accelerate the process of direct marketing by the farmers. Rural periodic market is the first contact point for producer - sellers for encashing their agricultural produce and buying other goods needed by them.

There are in all 27294 rural periodic markets including those for livestock, in the country. But even minimum necessary infrastructural facilities do not exist in most of these rural periodic markets. There is urgent need to develop these rural periodic markets in a phased manner with necessary infrastructure amenities to have a strong base level link in the marketing chain. Once developed, these places, where periodic markets function, can also serve as farmers/consumers markets. The investment requirement for developing these primary rural market places is estimated at ₹.2146 crores.

Storage/Warehousing

Storage infrastructure is necessary for carrying over the agricultural produce from production periods to consuming periods. Lack of adequate scientific storage facilities cause heavy losses to farmers in terms of huge wastage in quantity and quality of products in general and of fruit and vegetables in particular. Seasonal fluctuations in prices are aggravated in the absence of proper scientific storage facilities. Central and State Warehousing Corporations have constructed warehouses in the different States. Food Corporation of India and some State governments have also created warehousing facilities and godowns.

The total covered storage capacity available with FCI., CWC and SWC is estimated at 26.4 million tonnes. In addition, storage capaciaty of around 25.3 million tonnes is available with public, private and cooperative sectors. In the background of 200 million tonnes production the available storage capacity of 52 million tonnes is quite inadequate. It is estimated that about 20 million tons of grains are stored in the form of CAP(Covered and Plinth). This clearly shows that the country needs much more facility than what is available now. This is specially more important for hill and remote areas in several states. For an additional 20 million storage capacity the investment required @₹.2700 per tons is ₹.5400 crores. The private sector needs to be encouraged to enter the storage and warehousing activity and make investment of this magnitude.

The Finance Minister in the 2001-2002 Budget has announced creation of Rural Godowns for non-perishables on the lines of construction of cold storages under the back-ended subsidy scheme implemented by the National Horticulture Board. In this connection, in the light of past experience, the Expert Committee recommends that definition of rural godown should include a house, warehouse located in a rural area where a rural periodic market/market yard/sub-yard/collection centers for different agricultural commodities already exists. It may also include D class municipalities or harvesting centers. Any private entrepreneur, cooperative, APMCs, SWC/CWC, registered NGOs, Farmers Registered Organisations, Accredited microcredit organisations and shelf-help groups should be eligible to construct and operate rural godown under the scheme.

The Godowns should be of viable capacity and rat proof with scientific storage following the specification drawn by CWC/SWC. The growers/ process's, State procurement agencies, wholesalers, other agencies involved in PDS etc. should be included as potential users. The subsidy to the construction of godowns may be limited to 25 per cent of the total cost. The godowns should be declared as deemed warehouses under the State Warehousing. APMC market fee, sales tax, purchase tax, octroi etc. should not be leviable on the goods stored. Similarly, provisions of Essential Commodity Act, Labour Act, Mathadi Act, Shop Establishment Act, Industrial Disputes Act etc. should not be applicable to these Warehouses. The material stored in rural godowns should be backed by warehousing receipts with common regulatory framework for negotiability. The godown owners/operators should be permitted to play the role of on-lender so as to channel credit to potential users. Considering the importance of rural godowns to farmers, bankers/financiers should be allowed maximum spread of 2 per cent or less over the NABARD refinance rate.

Cold Storage

India produces 134.5 million tons of fruit and vegetables and the output is likely to go up during the next 10 years. It is a matter of concern that more than 30 per cent of fruits and vegetables produced in the country are lost due to lack of proper handling, stoarge and procesing facilities. Cold storages are most important infrastructural need for perishable and semi perishable commodities which need an immediate attention. Presently a total of 4199 cold storages are existing in the country with a total storage capacity of 15.38 million tonnes. The sector-wise availability of storage capacity is given below:

Table. Sector-Wise Distribution of Cold Storage Facilities (as on 31st March 2001)

S. No.	Sector	Number of cold storages	Capacity(Lakh MT)
1	Private	3739	146.13
2	Co-operative	310	6.80
3	Public	150	0.91
4	Total	4199	153.85

The present storage capacity available is sufficient only for 10 per cent of total production of fruits and vegetables. In the next 10 years with the anticipated increase in production of fruit and vegetables and other perishable commodities, the cold storage capacity requirement would be much higher. Foreseeing the future requirements of the fresh/precooked/ frozen fruits and vegetables and their products as well as anticipated change in the food habits in favour of processed food, the capacity requirement for post harvest management of perishables is estimated at more than five times the presently available capacity.

In the next 10 years, 15000 additional cold storage units with a capacity of 45 million tonnes should be created. The additional capacity requirement would need an investment of the order of ₹.27,000 Crores. The investment should basically be made by the private sector only. In future, there would be a need for multi-chamber type of cold storage units for various perishable and other products.

For encouraging private entrepreneurs there is a need to provide subsidy to make the units viable for some initial years. This apart, the regulatory arrangements should also be reviewed and simplified for attracting private investment in this venture. There is a need to provide incentives in reducing current expenses such as tax relief in electricity.

Reefer Vans/Containers

The country would also require reefer container/vans for transport of perishable items for domestic and export marketing. At present their availability in the country is negligible in comparison to the present production of perishable commodities. For handling the expected higher production in the next 10 years, at least 3000 reefer containers/vans with a capacity of 8 tons each would be required. This would require an investment of ₹.600 crores, which should be encouraged in the private and cooperative sector. There is a need to encourage the investors in this area by providing incentives.

Cleaning, Grading and Packaging

The importance of these facilities can be hardly over emphasized. At present, the grading facility is available only in 1321 markets out of total number of 7127 regulated markets.

The quantity graded at producers level is almost negligible. There is a need to create facilities for cleaning, grading and packaging not only at primary level but also in the villages from where produce is brought to the market for sale. In the absence of such facility at the village level, the kind of pollution and congestion created at market yards during the peak arrivals period is well known.

The APMCs should encourage private entrepreneurs to promote such units in or around the yard/sub-yards. There is need to promote proper packaging after grading so that further chances of adulteration or temptation may not be there. Besides this there is a strong need to educate the farmers for proper packaging and grading before they bring the produce to the market. Scientific packaging should be encouraged at the farm level through subsidy support. The Expert Committee feels that this is an important activity, and an investment of ₹.2000 crores should be earmarked for this purpose during the next 10 years.

Export Zones and Food Parks

With a view to taking advantage of new international trade environment, there is a need to encourage export of high value traditional/ non- traditional products grown in various parts of the country. Commodities having export potential are several fruits and vegetables, raw as well as processed and packed spices like cumin, fennel, coriander and other farm products like fenugreek and hena for which there is significant demand by Indian Diaspora and others in several countries. However, there is a need to educate and train the growers of these crops in producing, grading and packing for overseas markets and create necessary infrastructure.

A scheme of creating Export Oriented Agri-Zones (EOAZ) has been announced by the Govt. of India (Ministry of Commerce) which should be promoted by providing institutional and physical infrastructure in each of these as per the needs of the specific commodity. In some of EOAZs, there is also a need to establish what is called Food Parks.

In these parks, some common facilities like electricity and warehouse should be created with central government assistance which will help in attracting investment by the private sector and the state government. While most of the investment should be made by the private entrepreneurs, as a way of incentive, government should invest in common facilities, and quality certification. The estimated public investment is ₹.200 crores and private investment of around 400 crores on fifty such EOAZs. In identification of EOAZs and Food Parks, the Government of India, through APEDA and DMI should take an active stance, rather than leaving it to the state governments. The Committee further recommends that commodity wise export potential studies be commissioned before establishing EOAZs.

Processing and Value Addition

Considering the increase in demand for value added and processed products, there is a need to enhance the capacity of agro-processing sector.

This will not only help in stabilising the prices realised by farmers but also in creating employment in rural areas. The food-processing sector alone provides tremendous potential in this area. For attracting private initiative and investment in food processing, the Government of India through Department of Food Processing and National Horticulture Board have already formulated several schemes of assistance. A ten year tax holiday has been announced. However, the state governments should also come forward and grant relief in terms of sales tax and other local taxes on processed products. Cheaper processed products will expand demand for such products.

At present, value addition is estimated at only seven per cent and processing only two per cent of the total production. Within next ten years, there is a need to increase value addition to 35 per cent and processing to atleast 10 per cent. Quality control and standardisation will be extremely important in this endeavour. The Central government should establish or encourage a network of food analysis laboratories in the country. This will also be necessary to face competition from imported processed products.

The investment potential in value addition and food processing is quite large. According to our esimates, the potential is ₹.150,000 crores. If conducive policy environment and incentive frame work is created, private sector can be attracted to make investment of this magnitude.

Sources of Investment

The investment potential in agricultural marketing system is quite considerable. The investment in agricultural marketing system will go a long way in making agricultural sector vibrant and enable it to face the competition of liberalised international trade environment without adversely affecting the livelihoods of those who depend on farming. However, for realising the potential of investment, a major part of which need to come from the private sector, a conducive and favourable environment would have to be created. For attracting the private sector investment at a level visualised by the Expert Committee, there is a need for (a) making complementary investment by the State and Central government; (b) subsidising a few activities to enable the private sector initiatives to attain viability; (c) active stance by the Central government in some initiatives; (d) reducing the regulatory controls and simplifying the procedures; and (e) ensuring adequate credit flows to agricultural marketing activities.

It is in this context, that the Expert Committee recommends that agricultural marketing and trade activities like construction of storage structure; rural godowns; cold storages, reefer vans; cleaning, grading and packing houses; export oriented agri-zones; and agro - processing and value addition should be included under the definition of priority sector for the purpose of lending by financial institutions.

For a long term sustainable and integrated development of horticulture a fresh look is also required to be taken at some of the legal provisions as laid down under various acts and rules framed there under.

The following are important:

- *Excise Laws:* Presently the alcoholic beverages based on fruit and vegetables are clubbed with other alcoholic beverages. Whereas world over items like wines and beer which are based on fruit and vegetables and have low content of alcohol (ranging below 11-12 per cent) are considered as items of food and are promoted as health drinks. With removal of quantitative restrictions there is a possibility that larger quantity of such beverages will be imported into the country. This sector has not developed inIndia so far because the Excise laws have implications both in terms of financial pricing of the products as well as restrictive legal frame work for setting up and operation of units for the manufacture of these health drinks. The Government of India has given a boost to the industry based on fruits and vegetables by allowing a complete exemption to the products from Excise Duty, as per the declarations in the Budget for 2001-2002. However, this declaration covers the items like pickles, sauce, ketchup and juices, etc. It is proposed that a similar treatment be given to the alcoholic drinks based on fruit and vegetables such as beer, wine, tadi and apple-cider and other similar range of products.
- *Minor Forest Produce: Under Forest Act:* Presently a variety of products are being collected by petty contractors by ruthless exploitation from the forests. The policy should allow the production of such items on commercial basis on private lands. The forest laws should be amended to this effect. To cite some examples, the following fruits are regularly picked from the wild plants by the villagers and offered for sale in towns. These fruits fetch very good price and there is already trading worth lakhs of rupees in these fruits. So it is worthwhile to grow these fruits on regular scale as orchard or wasteland crops.
  - Wild pomegranate
  - Kaphal (Myrica nagi Thunb)
  - Wild fig
  - Lassora (Cordia oblique Wild).

Similarly, there are vast varieties of forest produce falling into the sub-categories of horticulture e.g. herbs, medicinal plants, vegetables, spices, flowers, honey, mushrooms, staple food supplements like buckwheat in the hills, etc:

- *Narcotic Drugs and Psychotropic Substances Act, 1985:* The present legal frame work is restrictive to the extent that it discourages the commercial production of the narcotic items which can be converted into pharmaceutical products. Therefore, provisions may be created for a balanced development for cultivation and processing of such items.
- *Plantations - A New Definition:* The word "Plantation Crops" as mentioned in some labour laws and Land Ceiling Act has a rather limited meaning pertaining mainly to coffee, tea, rubber, etc. To cover a large number of crops under plantations with a more rationale and more scientific approach is need of the day. If plantation is used as generic term for an advanced form of agriculture where various types of management from the selection of land, selection of species to be grown, the financial support, the management of labour, the processing and marketing are all done at a higher level than what is done for ordinary agricultural crops, the horticulture is bound to get a fillip. The existing legal framework will open new avenues for development of entire horticulture sector if production of fruit and vegetables, medicinal and aromatic plants, spices, is also brought under the definition of plantation.

Agricultural Marketing in Developing Countries

Economic reforms have had sweeping impacts on agricultural markets in developing countries.

In general, state intervention has been reduced, notably with respect to:

- The abolition or sharp curtailing of parasitical marketing boards;
- Depreciation of formerly over-valued currencies rendering developing country exports more competitive and imports more expensive;
- A reduced public role in agricultural services, especially in subsidized credit, input and extension networks;
- A shift away from pan-territorial and pan-seasonal crop pricing strategies and pre-announced prices.

  Reviewing agricultural markets research in sub-Saharan Africa and Asia, Jones concludes:
- "In newly liberalized [food] markets in eastern and southern Africa... barriers of entry to trade are low, but the marketing system has little capacity to channel credit or spread risk. There are strong

> theoretical reasons for expecting the impact of and response to reforms to vary between different classes of producers. The absence of key markets, risk aversion, high transaction costs and the dual role of agricultural households as producers and consumers are critical features. The marketing system depends on both physical and institutional infrastructure... Collective action by market participants may address this but it may also lead to collusion over prices. Evidence from South Asia shows that food markets exhibit social barriers to entry, massive asset polarization, debt relationships between large and small traders and traders and farmers, diverse institutional and contractual arrangements, and collusive behaviour, enforced in part by manipulation of the state regulatory system."

This conclusion gives some clue to the reasons why NGOs and CBOs intervene in agricultural markets. When extension agents, researchers and development organizations working in rural areas ask farmers to prioritize their problems, agricultural marketing is repeatedly raised as one of the most important problems faced. It may arise in the context of the promotion of new crops or productivity-enhancing technology, or it may be felt particularly acutely in remoter areas poorly served by commercial traders, where parastatals no longer operate.

NGO marketing interventions typically aim to fill critical gaps in the marketing system or address the power imbalances to which Jones refers. Nowhere are those marketing problems felt more acutely than in the areas for which it is most difficult to identify sustainable strategies to improve market access. Farmers in remote areas (either remote because of physical distance from markets or because of poor roads) are almost always poorly served by agricultural traders and are often obliged to accept seemingly unattractive prices for their produce.

Distance from markets rules out the production of higher value more perishable crops, and reduces the linkages between these producers and other more specialized markets. By the same token, CBOs and NGOs seeking to promote alternative strategies for these disadvantaged communities face high costs and tangible obstacles that make their task particularly difficult. Poor access to markets is mirrored by poor access to all kinds of rural services.

The poverty that results makes such communities particularly risk-averse. Where rainfall is uncertain, the situation is even worse, whilst the relative absence of trade does nothing to relieve the covariance in production. These are the challenging circumstances that make an examination of marketing interventions worthwhile. There is wide-ranging experience amongst the development NGO community. Some of these initiatives have taken-off and developed into self-sustaining activities, whilst others, although not conceived as such, have effectively become subsidy dependent welfare programmes. This review identifies best practice and the conditions required for such programmes to work.

NGOs and CBOs - Some Definitions

NGOs are part of the development landscape. Increasing amounts of development aid are channelled through NGOs.

The term gives little clue as to their real characteristics but most people associate NGOs with the following:

- A formal and officially recognized organization that is not linked to government;
- Having a purpose that is altruistic rather than commercial;
- Attracting staff who are value-driven rather than financially motivated.

Fowler highlights some features of the voluntary sector by making comparisons with government and business- organizations.

Table. Comparison of Organisations in Different Sectors

	Sector		
Characteristics	**Government**	**Business**	**Voluntary**
Relationship to those served based on:	Mutual obligation	Financial transaction	Personal commitment
Duration of relationship to those served:	Permanent	Momentary	Temporary
Approach to external environment:	Control and authority	Conditioning and isolation	Negotiation and integration
Resources from:	Citizens	Customers	Donors
Feedback on performance	(in) direct politics	Direct from market indicators	"constructed" from multiple users

The term 'community-based organization' or CBO may be associated with similar values but is generally more focused on issues particular relevant to the community from which its membership is drawn. CBOs may be quite formally structured, but can equally be loosely structured, informal organizations; farmers' associations are an example of a CBO.

Stocker and Barbor-Might discuss CBOs in relation to civil society organizations (CSOs): "CSOs are, simply, organizations operating in civil society somewhere between the informal associational world of family, kin, neighbours, friends and the more formal or market-oriented world of business organizations, state – and NGOs. Although in practice there is diversity and complexity, organizationally the idea is a simple one; NGOs act as intermediaries for CSOs...

CSOs sometimes evolve into NGOs. Many of them have nothing to do with development or are not poorer people's organizations. With this in mind, many authors and authorities prefer to use alternative terms, for example, 'GRO' (grass roots organization) or 'CBO' (community-based organizations) to designate CSOs that exist to serve their members, these members being poorer people. CBO is the usage followed by the World Bank."

Marketing functions of Agriculture

In modern marketing the agricultural produce has to undergo a series of transfers or exchanges from one hand to another before it finally reaches the consumer.

This is achieved through three marketing functions:

1. Assembling,
2. Preparation for consumption
3. Distribution.

Concentration pertains to the operations concerned with the assembly and transport from the field to a common assembly field or market. The produce may be taken direct to the market or it may be stored on the farm or in the village for varying periods before its transport. It may be sold as obtained from the field or may be cleaned, graded, processed and packed either by the farmer or village merchant before it is taken to the market. Some of the processing is done not because consumers desire it, but because it is necessary for the conservation of quality.

At the market the produce may be sold by the farmer direct to the consumer or more usually through a commission agent or a broker. It may also be purchased by traders, wholesalers or retailers. The transactions may be carried out by direct negotiation or through middlemen, by barter or cash, by open or under cover auction, on the spot or in future markets. The transactions take place at one or more levels in the primary, secondary or terminal markets or all three. Distribution (dispersion) involves the operations of wholesaling and retailing at various points. By a series of indispensable adjustments and equalising functions, it is the task of the distribution system to match the available supplies with the existing demand.

MARKETING FUNCTIONARIES (AGENCIES)

The transfer of produce or goods takes place through a chain of middlemen or agencies. In the primary market the main functionaries are the producer, the village or itinerary merchant, pre-harvest contractors, commission agents, transport agents etc. In the secondary market the processing and manufacturing agents are the additional functionaries. Financing agents such as shroffs, banks and co-operatives may also take part. In the terminal or export market the commercial analyst and shipping agent also gets involved in the transfer of goods.

The functionaries have their own set up. They may be individuals, partners or co-operatives who may buy and sell on ready and future basis at a price determined by forces of supply and demand. Each functionary renders some service in the process of marketing and also earns a varying margin of profit for himself. This procedure makes marketing rather complicated and inflates the price of the produce. The nature of some of the agricultural products for example their bulk form and perishability and their seasonal availability further add to the complexity of agricultural marketing.

MARKETING IMPROVEMENTS

India being a primary producing country, agriculture plays a vital role, both as an essential infrastructure and a development component in generating and sustaining a higher national income. Out of a national income of about ₹38,921 crores in 1972-73 as much as ₹17,500 crores or about 44.9 per cent is contributed by agriculture and allied sectors. It is estimated that about 50 per cent of the agricultural produce is available as marketable surplus.

The marketing system in India provides sustenance for about 3 million persons who are engaged in performing various marketing function. In the field of exports too, the agricultural sector accounts for about 50 per cent of the total value.

The process involved in the disposal of such a substantial produce of great economic importance are significant not only for the farmer but also for the country as a whole. The unreasonably low return that the farmer gets for his produce and the excessive margin of profit retained by the intermediaries attracted the Government's attention and it was felt that the economic condition of the agriculturists could not be improved unless determined steps were taken to establish an orderly system of marketing in the country.

GOVERNMENT REGULATORY PROGRAMMES

With this object in view, a number of marketing surveys were conducted by the Directorate of Marketing and Inspection which revealed the shortcomings in the country's marketing system. A rectification of these deficiencies was sought to be achieved by rationalizing various activities and standardizing various practices in the markets through legislation or otherwise. The primary objective of improving the system of agricultural marketing was not only to remove the handicaps from which the producer-seller was suffering but also to increase his income by ensuring him a fair price.

Regulation of Market

Prevailing market practices and market charges made a deep cut in the share of the producer in the price paid by the consumer. Some of the market charges were authorized whereas others were more than what the service rendered warranted. It was felt that a remunerative price to the producer could only be ensured if the market practices and market charges were regulated and rationalised. And thus the regulation of the markets has been given a high priority in the various Five Year Plans. The markets are sought to be regulated through an Act of each legislature. The Act is generally known as the Agricultural Produce Markets Act and it is provided for the removal of various malpractices widely prevalent in the markets for the settlement of disputes between sellers and buyers and for the promoting of orderly marketing of farm produce in general. Various state Governments have made considerable progress in this field by bringing in the necessary legislation. The Acts enabling the respective states to regulate the markets generally provide for the notification of market areas and the commodities to be covered in the act in different areas.

A 'Marketing Committee' consisting of the representatives of growers, traders, merchants, local bodies and Government nominees administers the working of each market. The functions of the market committee are to frame bye-laws, define local market prices, fix market prices payable to various functionaries, license the functionaries, settle disputes, supervise weightment and promote the development of orderly marketing in general. The committee is generally empowered to raise funds for its working by levying a small fee on the produce bought and sold in the market in addition to the license fees received from the functionaries.

Market Surveys

A survey conducted in 496 markets in 1961-62 has shown that after the regulation the market charges have been reduced by 48 per cent. Another survey of selected commodities revealed that for some commodities market charges have been reduced by as much as 98 per cent.

Although the first market to be regulated was Karanja in 1886 the regulation of markets did not make headway till the first Five Year plan. It got a fill up in the second and third five year plan.

The states and union territories which have regulated the markets are: Andhra Pradesh(335), Bihar(62), Chandigarh(1), Delhi(3), Gujarat(212),

Mysore(102), Madhya Pradesh(233), Haryana(80), Kerala(6), Maharashtra(212), Orissa(34), Punjab(95), Rajasthan(90), Tamil Nadu(136), Tripura(1), Uttar Pradesh(264), Goa, Daman and Diu(1), West Bengal(15), and Himachal Pradesh(5). The states and union territories which are yet to regulate markets are Assam, Andaman and Nicobar islands, Arunachal Pradesh, Dadra and Nagar Haveli, Jammu and Kashmir, Laccadive and Minicoy islands, Meghalaya, Mizoram, Nagaland, and Pondicherry. Necessary measures to regulate markets in these states and union territories are at various stages of progress. It is expected that these states and union territories will regulate the markets by the end of the Fifth plan.

All the states where necessary legislation has since been passed, have formulated phased programmes for the regulation of markets. By the end of the fifth plan it is expected that all the wholesale assembling markets would be brought within the regulatory orbit.

As a result of this scheme, excessive commissions and other market charges have been substantially rationalized. Unauthorized and arbitrary deductions have been prohibited and malpractices stopped. The issue of sale slips by licensed commission agents to the sellers, indicating the details of sale proceeds, deductions effected etc. has been made obligatory. Weightment is also done by licensed weightmen of the market committee.

The dissemination of marketing information and news is one of the functions of the market committees. This is done through the displaying of the prices prevailing in the market and also in the neighbouring markets on the notice boards and announcements through loud speakers at regular intervals. This information is also supplied to the Central Government and the State Government and also to other market committees. Arrangements have been made for the maintenance of reliable statistics arrivals, sales, stocks, prices etc. which are maintained by the market committees.

It is also obligatory on the part of the market committee to provide the market yards with the necessary amenities. In some of the markets that have been regulated amenities like rest houses, cattle sheds and water troughs have been provided by the market committees for the convenience of the producer sellers. Facilities for grading before sale and storage have also been provided.

A survey of 500 regulated markets was undertaken by the Directorate of Marketing and Inspection in the Ministry of Agriculture in 1970-71 and 1971-72, with a view to assessing the adequacy and efficiency of the existing regulated markets and highlighting their drawbacks and deficiencies and suggesting measures to develop them. One of the most important drawbacks has been the inadequate financial resources of some of the market committees. During the fourth plan, a central sector scheme was drawn up by the Ministry of Agriculture to provide a grant at 20 per cent of the cost of development of market, subject to a maximum of ₹ 2 lakhs. The balance will have to be provided by the commercial banks.

An important development in the field of regulated markets is the keen interest taken by the International Development Agency (IDA) in the development of the infrastructure in regulated markets. The IDA is financing the development of infrastructure in 50 markets of Bihar.

The World Bank has approved a loan assistance of 6.5 crores to Karnataka also for the development of markets.

Contract Terms

Under the existing trade practices, the sale of produce in a primary market takes place on the basis of the visual inspection of the goods, and in the secondary and terminal markets on the inspection of the samples. Thereafter the buyer and the seller decide upon the terms either orally or through written contracts. The contract terms specify the quality and quantity of the produce, the time and place of delivery, the price and terms of payment, handling and incidental charges, the procedure for settlement of disputes and penalties. The terms of contract were not standardized and thus varied for every individual transaction, and were more favourable to the buyer.

With a view to improving trade practices, All-India standard contract terms have been drawn up for a number of commodities. In standard contract terms the definition of quality and allowances in respect of refraction, damaged goods have been specifically standardized though the adoption of these standard contract terms by traders is voluntary they have to a large extent strengthened the position of the producer-seller and have improved the quality of the product marketed.

Standardization and Grading

In order to gain the confidence and establish a rational relationship between the quality of a produce and its price, it is necessary to devote some attention to the proper preparation sifting and sorting of a material according to certain attributes before it is taken to the market. This is sought to be achieved by grading the produce in conformity with certain accepted quality standards *viz.* shape, size, form, weight, and other physical and technical characteristics.

The produce brought to the market is very often contaminated with dust, stones and other foreign matter added either deliberately or by accident. Sometimes the produce is immature or not properly dried or contains shrivelled grains or damaged and rotten material. Such a produce brings a lower price to the farmers. Care should be exercised while assembling the produce of different farmers so that the good material is not mixed with the inferior material brought in by some farmers. The Government of India had recognized the need to introduce the standardization of agricultural produce which would enable the farmers to derive the benefits of grading in terms of fair practices according to the prescribed standards and enacted the Agricultural Produce Grading and Marking Act in 1937.

The Act empowers the central Government to prescribe grade standards indicating the quality of articles included in the schedule and specify grade designation marks to represent particular grades or qualities. The Act provides for the grading and marketing of agricultural produce. The grade standards prescribed under this act are based on both physical and chemical characteristics and are formulated after analysing representative samples of each commodity collected from different regions and different seasons.

Besides the international standards and special requirements of overseas consumers are also taken into account while formulating these standards for the commodities which are exported. The grade standards are reviewed and amended from time to time in the light of the shift of the pattern of production and trade and changes in the consumer's preferences. The grades are designated as the 'Agmark' grades.

A central Agmark Laboratory at Nagpur with sixteen regional laboratories at Guntur, Madras, Bombay, Kanpur, Cochin, Rajkot, Calcutta, Sahidabad, Jamnagar, Bangalore, Patna, Tuticorin, Virudhunagar, Mangalore, Alleppey and Kozhikode are assisting to provide adequate laboratory facilities for fixing grade standards for new commodities, for revising old grade standards and for routine quality control work.

Grade for Export

Grading of agricultural produce under the A.P. (G and M) Act is voluntary. Exports of certain agricultural commodities have however been prohibited unless duly graded and marked in accordance with the grade standards laid down under the A.P. Act 1937. The power to so prohibit exports was derived under the provisions of the sea-customs act. The Directorate of Marketing and Inspection under the Ministry of Agriculture exercises a three tier control on the quality of agricultural commodities that are graded under 'Agmark' before they are exported. This is done through inspection.

Grading for Internal Trade

Commodities such as cotton, ghee, butter, rice, wheat, atta, gur, eggs, arecanut, potatoes, fruits, bura, pulses, vegetable oils and ground spices are being presently graded under 'Agmark' on voluntary basis.

Grading at Farmers Level

The grading of agricultural commodities under 'Agmark' has been consumer oriented. Generally the grading was done at the level of the traders. At this stage the producer was not a direct beneficiary of the grading scheme. It was felt the need to introduce grading at the producers' level. Thus the Directorate of marketing and inspection introduced a scheme for setting up commercial grading units.

Grading of Fruits and Vegetable Products

With a view to exercising quality control over fruits and vegetables the Government promulgated the Fruits Product Order under the essential Commodities Act. The preservatives and colors to be used are also clearly laid down. The order also stipulates the hygienic and sanitary methods which must be adopted by the manufacturers. The license for all this is issued by the executive Director, food and Nutrition board.

The total number of licensed factories was 1194. The total average production of fruits and vegetable products in the country during 1970 has been estimated as 156 thousand tonnes valued at ₹32.66 crores. Some of the products are very popular in foreign markets and are a good source of foreign exchange. In 1970 alone fruit products worth ₹2.92 crores were exported from India.

Regulation of Cold Stores

Most of the problems relating to the marketing of fruits and vegetables can be traced to their perishability. Perishability is responsible for high marketing costs, market gluts, price fluctuations and other similar problems. At low temperature, perishability is considerably reduced and the shelf life is increased and thus the importance of cold storage or refrigeration.

The first cold store in India was reported to have been established in Calcutta in 1892. However significant progress in the expansion of the cold storage industry in the country has been made only after independence.

An ad-hoc survey of the cold stores carried out by the directorate of marketing and inspection in 1955 showed that the total available cold storage in the country was only 77 thousand tonnes. The survey also highlighted the need to regulate the cold storage industry in a planned manner.

With a view to ensuring the observance of proper conditions in the cold stores and to providing for development of the industry in a scientific manner, the Government of India and the ministry of agriculture promulgated an order known as "cold storage order, 1964" under Section 3 of the Essential Commodities Act, 1955. The order is applicable in respect of every cold store with a capacity more than 8.4 cu m. The jurisdiction of this Order extends to the whole of India except West Bengal. Under this order, it is obligatory on every operator of the cold store to obtain a license from the Licensing Officer before using the installation for storing any food stuffs *e.g.* fruit, vegetables, meat, fish, dairy products. The Agricultural Marketing Advisor to the Government of India is the Licensing Officer.

The directorate of marketing and inspection is enforcing the cold storage order. The field staff posted in the regional and sub offices located in different states regularly inspects the cold stores and offers necessary guidance for better and scientific preservation of foodstuffs. Besides the directorate of marketing and inspection gives general guidance on all technical matters concerning the setting up of cold stores to the intending entrepreneurs.

The Government of India constituted a Central Cold Storage Advisory Committee consisting of official and non-official members, representing the growers, owners, machinery manufacturers, research organizations etc. The Committee advises the Government on all matters pertaining to the enforcement of Cold Storage Order and the future development of the industry. At the end of 1973-74 there were 1503 cold stores in the country with a capacity of 18.70 lakh tonnes. Uttar Pradesh had the maximum number of cold stores of 9.37 lakh tonnes followed by Bihar with 169 cold stores with 2.10 lakh tonnes capacity and West Bengal with 133 cold stores with 3.15 lakh tonnes capacity. Out of 1217 cold stores licensed during 1970, 1021 representing 84 per cent of the total were owned by the private sector, whereas 121 and 75 were owned by the public and co-operative sector respectively. The respective capacity is 13.88 lakh tonnes, 0.25 lakh tonnes and 0.73 lakh tonnes. Though numerically the cold stores in the co-operative sector were less than those in the public sector the capacity was more. The National Co-operative Development Corporation (NCDC) has formulated a scheme for financial assistance for setting up new cold stores in the co-operative sector.

Potato is the most important commodity which is presently placed in the cold stores accounting for as much as 92 per cent of the total capacity in the country. The remaining 8 per cent is being utilized for other perishables *viz.* fruits, vegetables, meat, sea-foods, dairy and poultry products.

With the attainment of self-sufficiency in the production of food grains, greater attention is being paid to the increased production of protective foods such as fruits, vegetables, fish, poultry and dairy products. These products being highly perishable are required to be kept in cold stores. Owing to the inadequate cold storage facilities at present considerable losses occur in the case of these commodities. There is enough scope for expanding the cold-storage industry with a view to providing facilities for preserving and prolonging the shelf life of these protective foods which are essential for human health.

Consumer Protection

In the marketing process the producers and consumers are the two weak ends of the chain. It is incumbent on the part of the Government to protect the interests of both of them. Producers are sought to be protected through the regulation of the markets, grading at the producers level and other similar measures. Consumer's interests on the other hand are safeguarded by grading under 'Agmark' at the level of traders. The progress towards making the consumer quality conscious is slow. With the growing popularity of semi-processed foods the danger of sub-standard food articles being marketed has become manifold. With this in view steps have been taken in many directions, *e.g.* Acts relating to grading and standardization of agricultural commodities, certification marks in respect of manufactured goods, pure food laws, laws relating to weights and measures, and laws relating to the manufacture of fruits and vegetable products have been passed and enforced. Commodities such as ghee, vegetable oils, butter, honey and powdered spices are being graded and marked under 'Agmark' under the provisions of the Agricultural Produce (Grading and Marking) Act 1937. The Agmark attempts to provide a third party guarantee for the consumer. It not only certifies the purity of the product but also gives an indication of their quality by the grade-designating mark. The economic incentive to the producer-manufacturer is reflected in the premium the Agmarked product fetches in the market over the ungraded products.

Running parallel to the enforcement of laws, certain measures to complement the effort of achieving the overall objective of consumer protection are also being adopted. Monopoly procurement of food grains being operated by the Food corporation of India and other state departments achieves the two objectives of ensuring a remunerative price to the farmers and a uniform and reasonable price commensurate with the quality of the produce to the consumers.

The task of consumer education and protection is formidable and the Government machinery alone will be inadequate to accomplish this task. There is a need of active consumers' movement as in other countries. The Consumers' Guidance Society has been established with the objective of educating the consumers about their rights and responsibilities and advising them on the quality of products available in the market. Similarly it has advised the producers and manufacturers to abide by such standards as are necessary for the health and safety of the users.

Co-operative Marketing

The existing institutional structure of co-operative marketing is such that the co-operatives are functioning at the primary level, at the secondary level (taluka or district) and at the state level. In pursuance of a recommendation of the Dantwala Committee, efforts have been made to persuade the state Governments to divert the middle tier *viz.* the district marketing federations, of the functions legitimately falling within the purview of the state or primary marketing societies, so that a two-tier system can be brought into operation.

The co-operatives in different states have been federated into a central level federation *viz.* the National Agricultural Co-operative Marketing Federation (NAFED). At the end of 1960-70 as many as 3335 primary cooperative marketing societies were operating. Of these more than 500 were specialized commodity marketing societies for special crops, cotton, fruits and vegetables. In certain states intermediate organizations at district and state levels have also been established. At the end of 1960-70, 232 such societies were functioning.

They included some commodity federations also. At the state level 28 apex co-operative marketing federations are functioning. They are normally handling all the commodities. There are a few apex cooperative societies which are handling exclusively a particular commodity *e.g.* two apex societies in Gujarat are handling cotton, one is handling fruits and vegetables and one in Uttar Pradesh is handling sugarcane only.

In order to strengthen and develop co-operative marketing to an extent where it may have an impact on the marketing of agricultural produce, several measures have been initiated. Steps are also being taken to see that there is an effective co-ordination between the state cooperative departments and their counterparts dealing with agricultural marketing. Consequently there has been a significant expansion in the operations of marketing co-operatives.

This is exclusive of the value of agricultural requisites and consumer articles handled by the marketing cooperatives. The main commodities marketed by the cooperatives were sugarcane, cotton, oilseeds, fruits, vegetables and plantation crops.

In pursuance of the decision of the Government, a scheme of outright purchases of agricultural produce by the co-operative marketing societies was launched in 1964-65 in 200 selected marketing societies. The basic theme of this scheme has been to bring the small producers within the fold of co-operative marketing together for such farmers. To provide the necessary financial backing the scheme envisaged the creation of a price-fluctuation fund. The fund envisages meeting losses if suffered by the cooperative marketing societies at different levels as a result of outright purchases of agricultural produce.

The scheme has been operative in several states. During 1969-70 the value of the agricultural produce purchased under the outright purchase scheme was about 34 crores. As a result of this scheme there has been a greater involvement of cooperative societies in the marketing of agricultural produce.

Besides this scheme has given impetus to inter-state trade by the cooperatives. They transacted about ₹ 66.55 crores worth of agricultural produce during 1969-70.

Interstate Trade

The co-operative marketing societies are devoting increasing attention to interstate trade in agricultural produce. The main commodities are wheat, pulses, plantation crops, copra, spices, fruits and vegetables. The bulk of the transactions were made by the Punjab Apex Federation. During 1970-71 the NAFED acted as the agency of the co-operatives of Jammu and Kashmir for marketing their apples outside the state.

Co-operative Export of Agricultural Produce

The export of agricultural produce by the co-operative sector continues to be a growing activity. The bulk of the exports are made by National Agricultural Cooperative Marketing Federation Ltd., which accounted for exports worth ₹5.64 crores. Besides, the Gujarat State Cooperative Marketing Federation Ltd., the Khanna Cooperative Marketing Federation Ltd., the Kerala State Cooperative Marketing Federation Ltd., the Jalgaon district fruit sales societies and the coconut oil millers society also cooperated to exports to countries like Kuwait, Malaysia, Ceylon, Singapore, Bahrain, Doha, Dubai, Iran, Muscat. The Khanna Solvent Extraction Plant in the Punjab State exported de-oiled cake worth ₹. 35 lakhs during 1969-70. The main commodities exported by the cooperatives were pulses, chillies, onions, pepper, de-oiled cake, potatoes and kardi extraction meal. The exports were mainly made to Ceylon and other important markets were Mauritius, Kuwait, Doha, Bahrain, Hong Kong, the USSR, the UK, Iran, Czechoslovakia and France. Some of the traditional items of export have been marketed in non-traditional areas. Pulses were exported to Cuba and onions to Malaysia and Singapore. The cooperatives also assisted various agencies to export agricultural commodities. In this connection exports of coffee and raw sugar were made. During 1970-71, NAFED exported agricultural produce worth ₹5.26 crores. The Jalgaon District Fruit Sale Societies Cooperative Marketing Federation Ltd., directly exported bananas worth ₹34 lakhs to Kuwait and Bahrain Islands.

Co-operative Cold Stores

Co-operatives are also to facilitate the storage and marketing of perishable commodities especially seed potatoes. By the end of the Third plan the cooperatives had established 87 cold stores and the target for the fourth plan was set at 45 more. At the end of December 1971 there were as many as 96 cooperative cold stores with a capacity of 1.42 lakhs tonnes.

Agricultural and Livestock Marketing

Markets are the context, both physical and conceptual, where exchange takes place. Marketing includes all activities from the producer to the final including processing and distribution systems. The term producer includes farmers or pastoralists and the manufacturers of production inputs when they produce the commodity being marketed. The term consumer is used for anyone who is the final consumer of a product or the final user of a production input (*e.g.* pastoralists may consume butter and veterinary inputs).

The retailer is the final link in the chain from producer to consumer. Hence, an urban butcher is a retailer and so is a vaccinator in the government veterinary services who delivers a vaccination. The wholesaler delivers the product to the retailer. The term farm gate is the location of a sale where a farmer keeps his or her animals or produces his or her crop (*i.e.* on a farm in the case of settled cultivators or at an encampment in the case of pastoralists). The terms market actors and market agents are used interchangeably to represent any persons participating at any level of the market.

The objectives of marketing vary. For the individual producer or consumer, the objectives may be to maximize benefits from the resources available and to expand marketing operations in order to increase wealth. From a societal viewpoint, the objectives may be to encourage efficient allocation of resources, to create wealth and promote economic growth in order to improve the general welfare of society. Important considerations may also be to improve distribution of income between sectors of the economy and to maintain some stability of supply and demand for marketed goods. Agricultural marketing in Africa normally begins at the level of the individual smallholder. Producers usually carry out some or all of the marketing steps.

Often, because producers are also consumers, little of what is produced is marketed. Livestock owners may be only marginally market-oriented. Because of traditional attitudes towards wealth in cattle, owners may choose to hold cattle rather than market them.

Producers are likely to be some distance away from consumers. They may also be highly dispersed.

Both conditions affect the nature of the marketing and distribution process. Also affecting the process is the nature of agricultural and livestock commodities, which are rarely in consumable form when first entering the marketing system, and suffer from perishability or are otherwise susceptible to losses during storage/handling.

Agricultural, and particularly livestock products, are generally seasonal in supply and are more susceptible to natural shocks. In Africa, the marketing of agricultural products typically suffers from limited institutional support.

The role of marketing and trade in development

Marketing and trade play vital roles in the economic growth and overall development of a nation.

The major roles of marketing and trade in the national economy can be thought of in terms of:

- Specialisation in activities of comparative advantage
- Enhanced resource-use efficiency and trade
- Advances in marketing with economic growth.

Specialisation in Activities of Comparative Advantage

Without market facilities, areas must maintain diversified activities to produce their own food, shelter, tools and other needed goods. In the presence of a market, however, an individual can specialise in one activity and sell the surplus in order to purchase other needed goods. The individual is likely to specialise on the basis of acomparative advantage in that activity for which he or she has some special resource or ability. A comparative advantage exists when an individual or region can produce a good, relative to the price of other goods, more cheaply than another individual or region.

In livestock production, comparative advantage is often the result of agro-ecological conditions particular to a region making it suited to certain specialised activities.

The agro-ecological basis for production results in regional comparative advantage, whereby all of an area with that common agro-ecological base shares the ability to produce the good relatively more cheaply than another area.

Specialised activities lead to trade. The gains from trade will be the value of additional production made possible through specialisation and trade. The exact gains from trade will depend on the market prices of the goods with and without trade. This concept applies equally to individuals, who use their comparative advantage to specialise in one task, selling their products to trade for the other goods they need.

Enhanced Resource use Efficiency and Trade

Through specialisation and trade, a community is better able to utilise its limited resources. Specialisation and the resulting efficiency of resource-use is the basis for economic growth and development. As markets and economies develop, surpluses occur more frequently in profitable activities, creating new wealth, while products are moved greater distances than before. Thus, trade is a necessary ingredient for economic growth. Marketing is simply the means by which trade occurs.

Advances in Marketing with Economic Growth

As economic growth proceeds, several changes in marketing take place. With economic development, the activities and tasks of marketing increase. Activities such as storage and processing, packaging and retail distribution become more important. Greater activity moves away from the site of production and towards marketing. This, in turn, creates employment opportunities and further specialisation (diversification of the community). Since livestock products typically have positive income elasticities of demand, economic growth can lead directly to new opportunities for production. Thus, the livestock subsector increases in importance.

With development, more economic agents may enter trade, helping to improve marketing services and, in some cases, allowing the market to capture external economies of scale. This refers to a situation where the presence of many agents allows each one to operate at a lower cost. An example is the case where increased trade in some commodity (*e.g.* livestock allows for the establishment of large storage facilities (*e.g.* pre-slaughter holding areas), which lowers per unit storage costs. The physical infrastructure can also be affected in a positive way by large markets, in the form of better roads and communication, offering the potential for external economies of scale.

Requirements for market development

For market development to occur, rural areas must be effectively linked, in terms of information and infrastructure, through the middlemen in the marketing system with urban centres of consumption. With the shift in resources away from production to marketing services, small-scale processing can expand markets by increasing demand through diversification of the end products. Perhaps most important, and crucial to the reform of African marketing systems, is the requirement that the institutional and policy environments do not discourage or unnecessarily impede the actions of marketers. As well, property rights and contracts should be protected.

Another important factor in the development of markets is the disequilibrium between demand and supply. Producers and consumers then must exert greater efforts to cope with each others' requirements. Increased efficiency resulting from trade is not in itself sufficient to create wealth. A stable but static equilibrium, where supply meets demand, may no longer produce new wealth. Disequilibrium, along with technical and institutional changes, may be the conditions needed to move to even greater comparative advantage and efficiency levels.

Further, initial scarcity of resources (poverty trap) can cause subsistence activities to dominate, denying the surplus labour or resources necessary to invest in new knowledge or technology required to create comparative advantage. The institutional and organisational requirements necessary to expand markets may also be enormous. The role of property rights will play an important role, as marketing inherently involves transferring property rights. The nature of the society may restrict the scale on which such transfers can take place.

Agricultural marketing support

In the United States the Agricultural Marketing Service (AMS) is a division of USDA and has programmes for cotton, dairy, fruit and vegetable, livestock and seed, poultry, and tobacco. These programmes provide testing, standardisation, grading and market news services and oversee marketing agreements and orders, administer research and promotion programmes, and purchase commodities for federal food programmes. The AMS also enforces certain federal laws. USDA also provides support to the Agricultural Marketing Resource Center at Iowa State University and to Penn State University.

In the United Kingdom support for marketing of some commodities was provided before and after the Second World War by boards such as the Milk Marketing Board and the Egg Marketing Board, but these were closed down in the 1970s. As a colonial power Britain established marketing boards in many countries, particularly in Africa. Some continue to exist although many were closed down at the time of the introduction of structural adjustment measures in the 1990s.

In recent years several developing countries have established government-sponsored marketing or agribusiness units. South Africa, for example, started the National Agricultural Marketing Council (NAMC) as a response to the deregulation of the agriculture industry and closure of marketing boards in the country. India has the long-established National Institute of Agricultural Marketing (NIAM). These are primarily research and policy organisations, but other agencies provide facilitating services for marketing channels, such as the provision of infrastructure, market information and documentation support. Examples include the National Agricultural Marketing Development Corporation (NAMDEVCO) in Trinidad and Tobago and the New Guyana Marketing Corporation.

Several organisations provide support to developing countries to develop their agricultural marketing systems, including FAO's agricultural marketing unit and various donor organisations. There has also recently been considerable interest by NGOs to carry out activities to link farmers to markets.

Agricultural marketing development

Well-functioning marketing systems necessitate a strong private sector backed up by appropriate policy and legislative frameworks and effective government support services. Such services can include provision of market infrastructure, supply of market information (as done by USDA, for example), and agricultural extension services able to advise farmers on marketing. Training in marketing at all levels is also needed. One of many problems faced in agricultural marketing in developing countries is the latent hostility to the private sector and the lack of understanding of the role of the intermediary. For this reason "middleman" has become very much a pejorative word.

Agricultural Advisory Services and the Market

Promoting market orientation in agricultural advisory services aims to provide for the sustainable enhancement of the capabilities of the rural poor to enable them to benefit from agricultural markets and help them to adapt to factors which impact upon these. As a study by the Overseas Development Institute demonstrates, a value chain approach to advisory services indicates that the range of clients serviced should go beyond farmers to include input providers, producers, producer organisations and processors and traders.

Market Infrastructure

Efficient marketing infrastructure such as wholesale, retail and assembly markets and storage facilities is essential for cost-effective marketing, to minimize post-harvest losses and to reduce health risks. Markets play an important role in rural development, income generation, food security, developing rural-market linkages and gender issues. Planners need to be aware of how to design markets that meet a community's social and economic needs and how to choose a suitable site for a new market. In many cases sites are chosen that are inappropriate and result in under-use or even no use of the infrastructure constructed.

It is also not sufficient just to build a market: attention needs to be paid to how that market will be managed, operated and maintained. In most cases, where market improvements were only aimed at infrastructure upgrading and did not guarantee maintenance and management, most failed within a few years.

Rural assembly markets are located in production areas and primarily serve as places where farmers can meet with traders to sell their products. These may be occasional (perhaps weekly) markets, such as haat bazaars in India and Nepal, or permanent. Terminal wholesale markets are located in major metropolitan areas, where produce is finally channelled to consumers through trade between wholesalers and retailers, caterers, etc. The characteristics of wholesale markets have changed considerably as retailing changes in response to urban growth, the increasing role of supermarkets and increased consumer spending capacity. These changes require responses in the way in which traditional wholesale markets are organised and managed.

Retail marketing systems in western countries have broadly evolved from traditional street markets through to the modern hypermarket or out-of-town shopping center. In developing countries, there remains considerable scope to improve agricultural marketing by constructing new retail markets, despite the growth of supermarkets, although municipalities often view markets as sources of revenue rather than infrastructure requiring development. Effective regulation of markets is essential. Inside the market, both hygiene rules and revenue collection activities have to be enforced. Of equal importance, however, is the maintenance of order outside the market. Licensed traders in a market will not be willing to cooperate in raising standards if they face competition from unlicensed operators outside who do not pay any of the costs involved in providing a proper service.

Market Information

Efficient market information can be shown to have positive benefits for farmers and traders. Up-to-date information on prices and other market factors enables farmers to negotiate with traders and also facilitates spatial distribution of products from rural areas to towns and between markets. Most governments in developing countries have tried to provide market information services to farmers, but these have tended to experience problems of sustainability. Moreover, even when they function, the service provided is often insufficient to allow commercial decisions to be made because of time lags between data collection and dissemination. Modern communications technologies open up the possibility for market information services to improve information delivery through SMS on cell phones and the rapid growth of FM radio stations in many developing countries offers the possibility of more localised information services.

In the longer run, the internet may become an effective way of delivering information to farmers. However, problems associated with the cost and accuracy of data collection still remain to be addressed. Even when they have access to market information, farmers often require assistance in interpreting that information. For example, the market price quoted on the radio may refer to a wholesale selling price and farmers may have difficulty in translating this into a realistic price at their local assembly market.

Various attempts have been made in developing countries to introduce commercial market information services but these have largely been targeted at traders, commercial farmers or exporters. It is not easy to see how small, poor farmers can generate sufficient income for a commercial service to be profitable although in India a new service introduced by Thomson Reuters was reportedly used by over 100,000 farmers in its first year of operation. Esoko in West Africa attempts to subsidise the cost of such services to farmers by charging access to a more advanced feature set of mobile-based tools to businesses.

Marketing Training

Farmers frequently consider marketing as being their major problem. However, while they are able to identify such problems as poor prices, lack of transport and high post-harvest losses, they are often poorly equipped to identify potential solutions. Successful marketing requires learning new skills, new techniques and new ways of obtaining information. Extension officers working with ministries of agriculture or NGOs are often well-trained in horticultural production techniques but usually lack knowledge of marketing or post-harvest handling. Ways of helping them develop their knowledge of these areas, in order to be better able to advise farmers about market-oriented horticulture, need to be explored. While there is a range of generic guides and other training materials available from FAO and others, these should ideally be tailored to national circumstances to have maximum effect.

Enabling Environments

Agricultural marketing needs to be conducted within a supportive policy, legal, institutional, macro-economic, infrastructural and bureaucratic environment. Traders and others cannot make investments in a climate of arbitrary government policy changes, such as those that restrict imports and exports or internal produce movement. Those in business cannot function if their trading activities are hampered by excessive bureaucracy. Inappropriate law can distort and reduce the efficiency of the market, increase the costs of doing business and retard the development of a competitive private sector. Poor support institutions, such as agricultural extension services, municipalities that operate markets inefficiently and export promotion bodies, can be particularly damaging. Poor roads increase the cost of doing business, reduce payments to farmers and increase prices to consumers. Finally, the ever-present problem of corruption can seriously impact on agricultural marketing efficiency in many countries by increasing the transaction costs faced by those in the marketing chain.

Recent Developments

New marketing linkages between agribusiness, large retailers and farmers are gradually being developed, *e.g.*, through contract farming, group marketing and other forms ofcollective action. Donors and NGOs are paying increasing attention to ways of promoting direct linkages between farmers and buyers within a value chain context. More attention is now being paid to the development of regional markets (*e.g.*, East Africa) and to structured trading systems that should facilitate such developments The growth ofsupermarkets, particularly in Latin America and East and South East Asia, is having a significant impact on marketing channels for horticultural, dairy and livestock products.Nevertheless, "spot" markets will continue to be important for many years, necessitating attention to infrastructure improvement such as for retail and wholesale markets.

2

Farm Product Marketing Management

MANAGEMENT OF A FARM

Farm management is different from what is commonly confused with the work of a farm manager who manages a government farm as an agronomist. His function is normally limited to supervising and handling the day-to-day routine of a farm. It normally pertains to the existing pattern of the resource-use and crops-mix under which only the existing plan is executed, supervised and carried out. An intelligent farm manager may go a little further and look after the farm machinery to keep it going.

Farm management, is however, much more than that. Here we are not just concerned with the distribution of labour and irrigation water for day-to-day operations. The emphasis is on the decision-making function of evaluating and choosing between alternative strategies. A major concern is about adjustments which are more suitable and profitable and about exploring new situations and opportunities for maximization of income and satisfying other goals of a farmer. It is the approach under which the opportunity costs of the various resources are evaluated and adjustments in resource-use and enterprise mix are made to secure higher levels of farm income.

Following this approach, the emphasis on yield (productivity per unit of a resource, mainly land) is not ignored, but the greater focus is on the increasing of farm income through a second business organisation. As a business, farming requires the application of business methods and efficient management. To be an efficient manager, one must keep oneself abreast of developments in new technology, new practices, price trends, and economic outlook. Again a farm management man should identify the constraints in the external environment conditions, which hamper a farmer's opportunities and plans for making adjustments in his farm organisation and render the technically superior production plan economically unattractive to the farmer.

Farm Management and other Sciences

Basically farm management is economics in the context of fundamental definition of economics which involves three elements, *viz*. the scarcity of resources, their alternative uses and the objective of profit maximization. It is this science which deals with making rational, profitable and economic recommendations to the farmers. It also deals with the growth and stability aspects of economics.

The basic information about the multiple (alternative) uses to which the scarce resources can be allocated is supplied by other physical and biological sciences. The research in these sciences should continually generate the relevant data on alternative technologies and practices whose profitability can be tested under actual farm situations. And these data are required for the farm organisation as a whole and not far for a single hectare, animal or tree. The challenge to the farm management specialist is to be able to integrate and synthesize the diverse pieces of information from many disciplines such as agronomy, animal husbandry, horticulture, soil science, plant breeding, entomology, plant pathology, general economics, sociology and psychology, into an optimum 'package' which can be used on the farms with profit.

Agricultural scientists mainly put emphasis on the maximisation of yield rather than on the use of the optimum level of resources. But the goal in farming is not to make a profit on some single enterprise or from a part of the farm land as some *krishi pandits* do, but to use land, labour, and capital resources in such a way that they make the greatest contribution to the total profits from the entire farm. The superiority of this discipline, thus, lies in its treating the farm as an operational unit and tailoring the recommendations of all other disciplines to fit into an individual farmer's pattern of resources.

In the context of the recent technological breakthrough, management today should be viewed as a process within a rapidly moving frame of reference. "It is now more scientific, less artistic; more dynamic, less static; more versatile and less rigid". Farm management is forward-looking in its approach. Its task is not so much the improvement of the present farming practices but of the establishment of the whole sets of new production methods and farming systems which would put our agriculture on a continuously rising growth curve.

Government policies should change the economic environment to help the interests of the farmers to converge on the national goals. Studies on farm management to determine the responsiveness of the farmers to different levels of prices become extremely relevant in inducing the farmers to produce the quantities of different agricultural products and services needed by society. A rational pricing policy for water is needed. Canal water, for example is charged for without any direct relation to the quantity supplied. As a result, there is no incentive to the farmer to allocate water in the most economical manner. Farm management helps to identify such uneconomic practices and the most limiting factors. Irrigation or power (the bullock versus the tractor) may be a more important restriction than the limits imposed by land. In areas where water rather than land is the principal limiting factor and the marginal-value productivity of irrigation water is very high, it should be most carefully used rather than over irrigating a few hectares of crops needing high water consumption.

Good farm management can lead to a highly productive use of farm resources and can avail itself of the technological revolution now going on in our agriculture.

Application of Farm Management

As against the general contention that farm management is mostly applicable to crop farms, its application to other types of farming business, such as orchards, dairying, poultry, etc. is no less important. Faculty farm management has contributed more to citrus decline than the poor physical-chemical conditions of the soil, the stock-scion incompatibility, insect pests, disease and viruses. Intercropping with exhaustive crops seems to be one of the major reasons for the poor health of the plants. Citrus die-back or decline does not seem to be a disease by itself, but a mere symptomatic expression of physiological imbalance owing to the cumulative effect of several factors.

Similarly, if milk production has to be encouraged, it has to be made more profitable than cereal production. One has to demonstrate how the addition of milch animals to a farm fits into its resource organisation and yields higher returns.

With land as a scarce resource, efforts are made to intensify its use for maximising profits through multiple cropping. During a given period, the relative profitability of different crop rotations is evaluated and finally selected for adoption. 'Kufri Sinduri' variety of potato is not being adopted on a large scale in spite of its giving 10 per cent higher yield than some other varieties, *e.g.* 'Up-to-date'. The reason is simple. The former is a late variety and delays the sowing of wheat which decreases its yield. It is not 'Kufri Sinduri' potatoes versus 'Up-to-date' potatoes; rather, it is 'Kufri-Sinduri' followed by wheat versus 'Up-to-date' followed by wheat that counts in farm management.

Similarly, in West Bengal, where jute has been competing with paddy, the situation is changing with the introduction of high-yielding varieties of wheat, which can be rotated with paddy. The comparison in terms of the relative profitability now runs between jute and paddy plus wheat, and not just between jute and paddy. Owing to the paddy-wheat rotation becoming more profitable, jute production in West Bengal is suffering a set back. As Indian agriculture becomes more productive with the passage of time, the farmers will be confronted with the task of making the best selection from among many alternatives.

Scope of Farm Management on Small Farms

Sometimes it is said that management is not important on small farms. This notion is widely prevalent under those situations where the farms operate on a low level of technology. In the case of small farms, farm organisation cannot be changed as much as on a large farm, but the choice in respect of farm practices and methods of production, cropping intensity, etc. offer worthwhile alternatives. Japan's method of improving agriculture based on the principles of farm management clearly demonstrates that a small farm by itself should not be a hindrance to increased production and higher income. Small farms may have different problems, but the principles of the efficient use of available resources to obtain the maximum economic returns and family satisfaction remain the same in both the cases. The principles of farm management, since they deal with the allocation of resources, apply to small and large farms equally. A Chinese proverb is very appropriate to quote: "Although the size of a sparrow is small, yet it is physiologically as perfect as a big bird".

FARM INVESTMENT ANALYSIS

Once market prices have been determined for those items that enter the cost and benefit streams, this information must be arranged in "pattern" accounts to begin the assessment of the effects a proposed project will have on the farmers, public and private enterprises, and government agencies that will participate in project implementation. These accounts are central to the financial analysis of agricultural projects; they are always based on market prices. Although we will touch on the essential elements of financial analysis, much more could be taken up here. Just how elaborate the financial analysis must be for a particular project will depend on the complexity of the project. Most agricultural projects will call for a financial projection based on at least one pattern farm plan that is assumed for participating farmers. This pattern (or model) farm plan projects resource use and income flows for a group of similar farms participating in the project.

The financial projections for the private and public firms or project entities may be quite summary in nature for a simply organised project, but a project in which several different firms and project entities are concerned or one that poses special financial problems may involve a much more complex analysis. This will enable the project analyst to proceed with the confidence that he is preparing an acceptable financial analysis.

For more complex projects, especially those projects involving more complex public or private firms with specialised cost, revenue, or financing situations, the project analyst will have to move beyond what is discussed here. Agricultural project analysts may want to turn for more technical help to financial analysts or accountants, just as they would turn to agriculturalists or livestock technicians for their particular expertise. Many financial analysts and accountants, however, will not be familiar with methodologies of agricultural project analysis or with the particular analytical needs of the financial aspects of these projects, so even financial analysts and accountants may find that the following discussion may help them respond to a request for assistance.

Objectives of Financial Analysis

Six major objectives for financial analysis occur in analysing agricultural projects.

Assessment of Financial Impact

The most important objective of financial analysis is to assess the financial effects the project will have on farmers, public and private firms, government operating agencies, and any others who may be participating in it. This assessment is based on an analysis of each participant's current financial status and on a projection of his future financial performance as the project is implemented. Detailed financial projections are needed for this analysis.

Judgement of Efficient Resource use

The overall return of the project and the repayment of loans extended to individual enterprises are important indicators of the efficiency of resource use.

For management especially, overall return is important because managers must work within the market price framework they face. Farm investment analysis and financial ratio analysis provide the tools for this review.

Project analysts and others concerned with decisions on policies for national economic growth and development will have to look beyond the financial analysis-at market prices-and form a judgement about the effects of the project on real resources for the economy as a whole.

Assessment of Incentives

The financial analysis is of critical importance in assessing the incentives for farmers, managers, and owners (including governments) who will participate in the project. Will farm families have an incremental income large enough to compensate them for the additional effort and risk they will incur? Will private sector firms earn a sufficient return on their equity investment and borrowed resources to justify making the investment the project requires? For semipublic enterprises, will the return be sufficient for the enterprises to maintain a self-financing capability and to meet the financial objectives set out by the society?

Provision of a Sound Financing Plan

A principal objective of the financial analysis is to work out a plan that projects the financial situation and sources of funds of the various project participants and of the project itself. The financial plan provides a basis for determining the amount and timing of investment by farmers and for setting repayment terms and conditions for the credit extended to support the investment. It provides the same basis for an assessment of the investment plans and debt repayment capacities of public and private firms participating in the project. Finally, for the project as a whole, the financial plan is the basis for determining the amount and timing of outside financing-whether from the national treasury or from international sources-and for establishing how rapidly the borrowed resources should be repaid. The estimated effect of inflation on both revenues and costs should be taken into account in making this assessment.

Coordination of Financial Contributions

The financial plan allows the coordination of the financial contributions of the various project participants. The coordination is made on the basis of an overall financial projection for the project as a whole. It addresses itself to such questions as whether the availability of resources from the treasury or international agency is matched with farmers' investment capacities and available funds for investment and operating expenses as well as with the timing of expenditures for project investments such as feeder roads and irrigation structures and for working capital needed for stocks in processing industries and the like.

Farm Planning

Planning means taking decisions in advance. It stimulates thinking, broadens understanding and challenges the farmer to move forward. It is a forward-looking approach.

The farm plan helps a farmer to decide how to combine new ideas and old ones to his best advantage. By identifying his credit and supply needs, the farm plan helps him to arrange for the timely supplies of credit, seeds, fertilizers, etc. A specific farm plan setting fort his expected output, expenses and income, serves as a sound basis on which a credit institution can give him production credit, based on his productive capability rather than on his net financial assets. It is out of his income and not through the sale of assets that the cultivator has to pay off his loan. Thus the farm plan or the budget is to the farmer what the blue-print of the architect is to a building contractor. It shows what is to be done and how it is to be done. It furnishes an organised and logical approach to his problems and helps him to work out the solution.

Different Types of Farm Budgeting and Planning

There are three main types of farm budgetings':

1. Enterprise budgeting
2. Partial budgeting
3. Full or complete budgeting and planning.

Enterprise Budgeting

The enterprise budgets are the input-output relationship for individual enterprises. An enterprise budget includes all the variable resources required per unit (a hectare/animal/tree, etc.) of an enterprise and its cost, the expected output, gross returns, etc. A format of an enterprise budget is given in Table.

Table. Format of an Enterprise Budget

Items	Quantity(kg/litre)	Value(₹)
Gross Returns		
Main product		
By-product		
Total		
Cash Variable Expenses		
1. Seed and seed treatment		
(i) Seed		
(ii) Sub-total		
2. Manures and fertilizers		
A. Farmyard manure		
B. Chemical fertilizers		
(i) CAN urea		
(ii) Superphosphate		
(iii)Muriate of Potash		
Sub-total		
3. Insecticides and fungicides		
4. Tractor fuel cost		
5. Irrigation hours		
6. Human labour		
7. Threshing hours with diesel engine		
8. Cost of typing material		
9. Marketing charges		
Sub-total		

10. Interest on variable expenses for half the period of growth
11. Total variable expenses
12. Returns over variable expenses
13. Man-hours
14. Bullock labour (pair hours)
15. Machine (tractor hours)

Enterprise budgets provide useful information regarding the resources requirements and the relative profitability of different enterprises. Thus these budgets, considered in the framework of farm resources, are the alternatives from among which the most profitable ones are to be selected.

In this context, the enterprise budgets need to be prepared at different levels of technology:

- The existing level of technology;
- The improved or recommended level of technology.

A comparison of the enterprise budgets at the existing and improved levels of technology provides the scope or the potential of making farm improvements.

The enterprise budgets lack in one important aspect that these do not consider the complementary and supplementary relationships amongst themselves which are quite common among farm enterprises at low level of production, but they simply assume to be competitive to one another right from the beginning.

But these relationships are taken care of in complete planning and budgeting.

Partial Budgeting

Partial budgeting is a method of making a comparative study of the cost-and-return analysis resulting from a change in a part of the business organisation.

This change may be made through a careful selection from among alternative methods of production or practices, the choice of which is based on the opportunity cost of relative profitability and does not affect the total farm organisation vitally. This technique helps to make decisions whenever small changes in the existing farm organisations are contemplated.

The following four points are important in setting up a partial budget:

- Additional returns from change
- Reduction in unit cost
- Reduction in yield, if any
- Addition in cost incurred

Put in a format, a partial budget will look like this:

Debts		Credit	
	₹		₹
(a) Increase in costs		(a) Decrease in costs	
(b) Decrease in returns		(b) Increase in returns	
Gain		Loss	
Total		Total	

Thus partial budgets deal with such changes in the farm organisation as can increase farm incomes without changing the total farm organisation. The farmer would know the total net benefit from the change, the complete details of what he should do at what cost and what he is not to do after the change and come out with higher profits.

Where Partial Budgeting is Used

Partial budgeting maybe used where the change in the activity under study would not affect the farm organisation vitally.

It has wide application and can be used to answer a wide variety of questions such as:

- The economic combination of CAN and superphosphate versus diammonium phosphate with a basal dose of CAN,
- The mechanical thinning of fruits like plums and grapes versus chemical thinning,
- The substituting of a more economical dairy ration for a costlier one (a single or two nutrients), and
- American cotton versus hybrid maize versus desi cotton, etc.

Partial Budgeting Distinguished from Full Budgeting

1. Partial budgeting considers a few alternatives which do not affect the organisation vitally, but full budgeting takes care of all the alternatives.
2. In full budgeting, the inventory of the farm, the resource structure, the existing resource use and such problems as overstocking or under stocking of resources, etc. are considered, but in partial budgeting information with respect to a few alternatives is considered.
3. Partial budgeting does not indicate the break-even point as to when to start one practice and abandon another, but full budgeting does.

The best strategy is to make an effective use of the partial budget at different stages of full budgeting.

Limitations of Partial Budgeting

The partial budgeting analysis suffers from some limitations, *viz.* it does not consider all the alternatives open to a farmer within the restraints of his present resources. It only considers a few alternatives and these too, should not affect the farm organisation to a great extent.

Partial budgeting does not always provide a complete solution. As one may demonstrate that mechanical threshing costs less per quintal than the traditional *phalla* method, but this would not serve the purpose. The other questions such as the break-even point of the thresher, *i.e.* the minimum volume of business (*i.e.* 1,000 quintals of wheat to thresh) the possibilities of custom work, the capital available to purchase the thresher, etc. should also be considered. Similarly, the question whether to increase the number of dairy animals may be answered by using partial budgeting. But, in this case the partial budgeting analysis will approach very closely to a change in the complete farm organisation. Again, the partial budget may show American cotton to be less profitable than hybrid maize, but when full budget is prepared, both the activities may not enter the farm organisation.

Like any other farm-management analysis, the result of partial budgets are subject to variations in the input-output prices and field conditions such as soil type, soil fertility, etc. Also the restraints pertinent to a particular farm organisation necessitate the tailoring of these results to ensure their applicability to means for making profits.

Complete Planning and Budgeting

In preparing complete plans for the farm, all the sophisticated analysis of studying an individual cultivator's opportunities, constraints and problems is done. Every little aspect of the farm organisation is examined and then suitable adjustments are studied and a suitable combination as fits in rationally with that given farm organisation is suggested. Various recommendations in terms of augmenting resources, if the need may arise, are also made. It implies the following steps.

1. *Farm Map*: The farm is carefully mapped out, giving its salient features, like soil type soil-fertility and rotations followed. Low-lying areas or other such features are also shown in the map. Then based on the previous crop history, land-capability classification is done and is also shown in the map.
2. *Inventory of Farm Resources*: Every asset on the farm, ranging from hand-tools to sources of power, etc. are inventoried. It does not provide us with the picture of the resources as owned by a farmer, but we can also work out their use-patterns and their condition,

i.e. whether they will have to be replaced or whether they will be sufficient for the new plan or some augmentation will be needed.

3. *Examining the Existing Organisation*: Having prepared an inventory of the existing resources and their availability, what we are interested in is their use-pattern within the framework of the existing crop mix, whether the resources are under stocked or overstocked. A careful analysis of the restrictions and weaknesses of the farm organisation is made. The weaknesses maybe many, such as:
   i. Lumpy units may not be fully utilised.
   ii. Irrigation may not be adequate.
   iii. The peak-season labour requirements may not be met.
   iv. Short-term capital may be inadequate.

   It is not simply the weaknesses that are emphasised but the good points of the existing organisation are also to be highlighted. The existing organisation may be improved or new good items maybe adopted.
4. Laying Down Restrictions and Planning.
   (a) Restrictions maybe with regard to bullock-power availability, or in respect of putting some area under cotton, or vegetables for home-consumption, though such a change in the cropping pattern may not be profitable.
   (b) Labour-requirements, power requirements, capital needs and new equipment needed are worked out and a suitable crop mix is adopted.

In complete farm planning, the goals of the farmer are also important. He maybe desirous of investing in farming and, at the same time, he maybe willing to construct a new house. His goals and opinions and risks are all studied and weighed, and a compromising decision maybe taken. Thus in a complete farm-planning, due consideration is payed to resource use, restrictions, importance of relationships among different enterprises and to the goals and managerial skills of the operator.

Analysis for Credit

A banker's plan is something more than the farm management specialist's plan. A banker is more concerned with the recovery of the loan advanced by him to the farmer. A farm-management specialist's plan is all right, but a banker, if he is to call it his plan, must examine it, as not only the additional returns from the alternative plan, but also whether the farmer would be able to repay the loan from the additional returns or not.

The credit forms the total capital requirement minus the farmer's owned funds. Most of the banks expect the farmer to contribute 25 per cent of the total capital requirement as margin money so that he has some stake in the business.

Season-wise Analysis

Because the loan is to be paid at the end of every season, repayments include:

(a) The total amount in the case of self-liquidating loans which are for short periods and for purchasing seeds, fertilizers, hiring casual labour, etc.

(b) The installments of the medium and long-term loans which are not self-liquidating and are for constructing buildings, raising irrigation structures, purchasing implements, machinery etc.

The format to compare the existing and the alternative plans for credit is given in Table. Section A gives crop combinations which can increase yields and farm income in an alternative plan. Section B shows how to work out short-term credit requirements. Section C shows how to work out the repaying capacity and economic feasibility of the credit proposal.

Table. Format to Compare Existing and Alternative Plans for credit

(A) Cropping pattern	Existing plan (without credit)	Existing plan (without credit)	Existing plan (without credit)		Alternative plan (without credit)	Alternative plan (without credit)	Alternative plan (without credit)
Season/ crop	Area	Yield hectare	Total production	Area hectare	Yield production	Total production	Gross returns
Kharif							
1							
2							
Sub-total:							
Rabi							
1							
2							
Sub-total							
Total							

(B) Short term credit requirements	Capital and credit requirements (Existing plan)	Capital and credit requirements (Existing plan)	Alternative plan	Alternative plan
Item	Kharif	Rabi	Kharif	Rabi
I. Variable expenses				
(i) Seed				
(ii) Fertilizers/manures				
(iii) Insecticides/fungicides				
(iv) Hired casual labour				
(v) Diesel petrol				
(vi) Miscellaneous cash expenses				
Sub-total:				

II. Fixed expenses
- (i) Land revenue/ water-rates/cash rent
- (ii) Permanent labour
- (iii) Flat-rate electricity bills
- (iv) Repairs

Other miscellaneous expenses

Sub-total
Total operational expenses (variable and fixed)
Farmer's availability
Short-term credit
Interest on credit amount
Amount to be repaid

(C) Economic feasibility tests of the farm credit proposal

I. Return analysis

	Existing plan (Kharif)	Existing plan (Rabi)	Alternative plan (Kharif)	Alternative Plan(Rabi)
(i) Gross income(A)				
(ii) Total operational expenses(B)				
(iii) Returns from fixed factors				
Net farm income:				
Incremental income:				
II. Repayment capacity:				
(i) Returns from fixed factors				
(ii) Income from other sources				
(iii) Total family income(i plus ii)				
(iv) Old debts/loan installment(s)				
(v) Family living expenses				
(vi) Social expenses (unavoidable)				
(vii) (iv plus v plus vi)	Total deductions			
Repayment capacity				

(iii minus vii)

Viii Risk-bearing ability:

(ix) Less% of income against risk and uncertainties

(x) Net repayment capacity(vii minus ix)

Farm-Efficiency Measures

Efficiency is the ratio of the output to the input. This concept is important as it shows how much profitable the farming business is. Various measures to explain the efficiency of the business as a whole, and in parts, are available. When examined together they help to point out the weaknesses in the farm business and provide a guideline as to which part of the business deserves special attention for making improvements. In a particular situation due importance is given to a particular measure. For instance different measures would be adopted for indicating the volume or size of the business, aggregate earnings from particular factors or the business as a whole and returns per unit of a particular factor input.

Further, the efficiency of a farm can be judged from the costs or returns or both. No single efficiency measure is so complete as to give a true picture of the entire farm business. These efficiency measures help to reorganise the same farm for which they are calculated but they must be used with great caution, while comparing different farms. This is especially important in the developing countries where farming is of diversified nature and individual farm business has wide variations in respect of soil type, resource restraints, capacities and capabilities of the farmers to undertake risks, their attitude towards innovations, etc. Also, these measures suffer from limitations for making comparisons because of their changing prices, costs, and conditions of the farm business. Adjustments with the price and coat indices are necessary before such comparisons give any valuable information.

Farm Labour

The hired farm labour comprised $21.9 billion or 9.1 per cent of total production expenses. That was an increase in monetary expenses of $3.3 billion, compared to $18.6 billion in 2002, but a decrease in percentage of expenses. In addition, contract labour amounted to $4.5 billion in expenses or 1.9 per cent of total production expenses, up $1.1 billion from 2002. Custom work and custom hauling, which includes machinery costs was up by $0.8 billion at $4.1 billion; 1.7 per cent of total production expenses (2007 Census of Agriculture). Hired labour was the third largest expense group behind purchased feed and purchased livestock and poultry. But farm labour expenses are not equally distributed regionally. According to Kandel, total farm labour expenses amounted to 22.3 per cent of the cash receipts in California, but only to 2.5 per cent in Iowa in 2006. The top five states in terms of payroll expenses were California, Florida, Texas, Washington, and Oregon. They account for 42.8 per cent of the expenditure on hired labour in the U.S. Runyan reported that 1910-19 the share of family labour of total farm employment was 75 per cent; 1990-99 this share had declined to 64 per cent. While total farm employment is declining, the role of hired workers is increasing with increasing farm sizes. However, farm wages rank near the bottom of all occupational groups, second only to private household work (Runyan). This fact may be ameliorated, at least in part, by lower cost of living expenses in rural communities (Gisser and Davila). By agricultural specialisation, hired labour is most important for horticultural operations (tree nurseries, ornamentals, fruit, and vegetables) and in dairy farming, followed by livestock and poultry farming; hired labour is least important in field crops.

Objectives of personnel management research in agriculture and agribusiness. The foci and results of personnel management research in agriculture and agribusiness in the United States and in Canada, but results are likely applicable beyond these two countries. The goal of the review is to extract the lessons learned and derive guidance for both agribusiness management practice and future research. The specific objectives are to

- Analyse the state of the art of personnel management research in agribusiness, in particular agricultural production, including an analysis of research methods;
- Determine the main themes with respect to
  - Research questions and
  - Empirical fields; and
- Summarize empirical results to
  - Provide a foundation for manager training and decision support and
  - Serve as a roadmap to future research projects.

Procedures

Geographically, this chapter focuses on the United States and Canada and the review is limited to publications in English. The analysis concentrates on publications analysing personnel management questions, largely excluding labour market, migration, immigration, and similar analyses. Labour market, migration, and immigration studies are important to understanding the agricultural labour problem and a considerable amount of work has been done on these questions. Less work has been published on personnel management functions and the use of different management practices in agribusiness.

Personnel management functions include practices to recruit, train, manage, organise, evaluate, compensate, discipline, and terminate employees, as well as, questions of job satisfaction, motivation, and retention. Therefore, the unit of analysis is the agribusiness or farm, not the market, society, or other macro institution. The review covers agribusiness and management journals, agricultural economics journals, and also animal science and horticultural science journals. In addition, research reports and conference papers (gray publications) are included when accessible and relevant.

Articles reporting on empirical research, as well as, review articles were content analysed with respect to the objectives outlined above. A qualitative analysis method was used to determine the personnel management questions addressed, the research methods, the empirical field, the specific results with respect to the questions addressed, and the broader implications of each article. Only articles meeting the criteria summarised above are included in the discussion of the main themes and in the summary tables. Furthermore, although this chapter is based on a comprehensive review, it cannot claim to include every study in this field.

State of the Art

Before 1990, personnel management was virtually absent from agribusiness and agricultural economics research (Howard and McEwan; Rosenberg and Cowen), with very few exceptions (*e.g.*, Adams, How, and Larson). For the agricultural field, personnel management research basically began in the early 1990, but many of these studies are difficult to access, because they have been published as conference or working papers, or in trade magazines, not in peer reviewed journals.

Until the end of the 1990, studies remain few and common themes are yet to develop, with the possible exception of job attitudes, which appear as an early focus (*e.g.*, Adams, How, and Larson). Additional themes emerging later include managers' conceptualisation of the personnel management functions, managers' personnel management competencies and practices, and the relationship between personnel management practices and organisational outcomes.

Few studies focus on one particular personnel management function; more studies encompass a broad array of functions and the related management practices. Exceptions are studies of the management and preferences of migrant workers and of compensation. Compensation studies in agribusiness frequently are limited to a description of actual wages and their distribution, sometimes not including benefits, and not relating compensation to organisational outcome variables. Examples of compensation studies, which transcend this limitation, are a pay method and performance study (Billikopf and Norton), a study of the effect of compensation and working conditions on retention (Gabbard and Perloff), and studies of the relationship between wage, production technologies, and farm size. Gabbard and Perloff found that for the same monetary investment employee benefits increase the probability of retaining good workers more than higher wages. Strochlic et al. also found benefits to increase retention. No relationship between wages and retention rates was found by Miklavcic, as well as Strochlic et al.

Considering that, regardless of the personnel management model used, specific management practices cannot be considered to function in isolation and independent of other practices. Conclusions based on such studies of singular practices would be limited. Therefore, even researchers interested in a particular personnel management function and in comparing relevant practices for this function, would have to take a more integrative approach and describe other practices to provide context. Empirical evidence for the relevance of the integrative approach in agriculture and agribusiness was provided in Adams, How, and Larson; Chacko, Wacker, and Asar; and Mugera and Bitsch.

Despite many commonalities between different branches of agricultural production, the type and conditions of work vary, as does the dependency on weather and growth cycles, *e.g.*, comparing vegetable production to swine production. Both researchers and practitioners therefore will primarily look at the research matching their current undertaking most closely. Studies vary in their empirical coverage, with respect to the scope of farming specialisations included, from studies focused on a single specialisation (*e.g.*, floriculture) to studies including multiple specialisations (*e.g.*, horticulture, including floriculture, fruit and vegetable production), and the scope of personnel management functions analysed, from single function studies to studies including selected or multiple functions. Dairy farming stands out as the specialisation most likely to be researched. Given that hired labour plays an even larger role in horticultural production than in dairy farming, the reasons for the higher interest in personnel management in dairy research are not obvious.

The Journal of Dairy Science published papers of a Symposium: Dairy Personnel Management as early as 1993. In addition to the dairy studies, that address personnel management specifically other studies of dairy farming included personnel management questions in broader studies of farm expansion. These studies found that personnel management competencies are most important for the success of farm expansion, but these competencies are also most challenging for farm managers. After an expansion, managers are more likely to use formal practices with respect to all major personnel management functions (Stahl et al.), but some problems, such as communication, persist (Hadley et al.), although managers spend more time on personnel management. Also, personnel management education for large dairy farms has been emphasized as an opportunity for extension programming (Brasier et al.).

A relatively new arena of research, which cuts across different agricultural specialisations, is the interface of personal management and sustainable or organic production. The questions being asked include whether sustainable and organic agriculture are inherently beneficial to employees, whether the commitment to sustainability does or should include a social component, and whether a fair labour certification approach would be beneficial to producers. Although a majority of certified organic farmers in California believe that organic agriculture is more socially sustainable than conventional agriculture, there is little support to include criteria on working conditions in the organic certification (Shreck, Getz, and Feenstra). On the other hand, Strochlic et al. found considerable interest in a fair labour certification (59 per cent of respondents).

Research Methods of Empirical Studies

Considering the early stage of personnel management research in agribusiness, research methods were expected to be mostly exploratory and qualitative. However, research methods cover the full range from in-depth, unstructured interviews and group discussions, through interview or moderator guide based approaches, up to fully structured surveys administered at the business site or off-site one-on-one or in a group setting, over the phone, or mailed questionnaires. Fornaciari and Dean found a similar phenomenon in the study of religion, spirituality, and management, where research methods also include many quantitative approaches, despite the early stage of the research field. Reasons for the seemingly early venture into highly structured and quantitative research approaches are more likely to be caused by expectations set up in the qualification process of researchers, professional pressures regarding publication outlets, and differing prestige of certain research approaches in researchers' professional fields than by research considerations.

Although, this review of studies of personnel management in production agriculture and agribusiness cannot claim completeness, the number of studies employing unstructured or moderately structured methods appears lower than the number of studies employing highly or very highly structured methods.

However, even many of the quantitative, highly structured studies did not attempt (or accomplish) representative sampling and, therefore, their generalizability can only be judged based on their descriptions of the research approaches and the methods used, and the comparison of results across studies. As a result, researchers and practitioners planning to use studies of either research approach may need to analyse the original sources and pay close attention to details, before evaluating the applicability of their results to a different context.

Most studies rely on a single method for data collection and multi-method studies are rare. An exception is the case study approach of best management practices by Strochlic and Hamerschlag that employed a variety of methods including semi-structured interviews with farm managers, focus groups with employees, and informal interviews with key informants. Multi-method approaches are likely to yield more valid results, due to the method triangulation involved.

The method used most often by personnel management researchers in production agriculture and agribusiness is a survey questionnaire. Questionnaires are administered in a variety of ways, most frequently in person, which is more likely to garner to reliable results than mailed questionnaires, given the sensitivity of many personnel management questions, but also requires more resources. The number of studies using a mailed questionnaire is surprisingly high, considering the difficulty of developing a questionnaire that is fully understood by potential research participants. Other methods used frequently are moderately structured interviews either in an individual setting or set up as group discussions. Although resource intensive, these latter approaches are more likely to gather reliable data and allow for in-depth study of research questions than the more highly structured approaches, given the early stage of the field, the lack of common understanding of personnel management terms of potential research participants and researchers, and the multitude of interactions between personnel management practices.

Managers' Conceptualisation of Personnel Management Functions

As early as 1967, Adams, How, and Larson observed that some farmers seem to have much fewer difficulties in finding and keeping the workforce they needed than other farmers in a comparable situation. Their research showed that this difference was not a chance occurrence, but that these farmers had invested considerably in the relationships with their workforce and carefully developed their personnel management practices. Similarly, Rosenberg and Cowen found dairy managers' assumptions about their workforce to correlate with their milk output, and suggest that those assumptions guide the choice of organisational structure and the management practices. Hence, it may be concluded that managers' perception of which personnel management functions need to be given attention and which practices are available to them, will be the determinants of their management choices.

After 2000, a renewed effort to delineate the field of agricultural personnel management resulted in three studies using focus group discussions to identify management practices in different areas of agricultural production and services, to describe their advantages and drawbacks from managers' perspective and to critically review these practices. As a research method, focus group discussions are useful to integrate research and extension goals. The interaction between research participants and between research participants and the researchers triggers learning processes. In addition, relationships are developed and reinforced, which not only increase openness during the research process, but encourage participation in educational programmes. During the research process, knowledge deficits can be diagnosed.

Bitsch und Harsh convened five focus groups with managers and owners of greenhouses, tree nursery operations, and landscape operations in Michigan. The study showed that horticultural managers conceptualise personnel management and its challenges and opportunities along the management process: recruiting and selection, training and development, performance appraisal and discipline, careers and relationships, and compensation. For the research participants, hiring immigrants and labour legislation were also important HRM topics.

In addition, Bitsch et al. convened four focus groups with dairy farmers and managers. Their perceptions of personnel management functions were similar to the horticultural study, and differences were mostly due to the more seasonal character of labour needs in the earlier study. Discipline was more important in dairy farming, because the continuous availability of work creates the need for terminating and replacing some employees who do not perform at the expected level. Seasonal operations often deal with these employees by providing less work to them, laying them off before the end of the season, and not recalling them for the following season. While horticultural managers considered working conditions mostly as an image problem in recruiting, to dairy managers working conditions were a permanent stress on employees.

Labour laws and regulations were less important in dairy farming, because few operations had their practices audited by government agencies at the time of the study.

Finally, Bitsch and Olynk (2008) convened six focus groups with owners and managers of pork farms in Kansas and Michigan and reanalysed the transcripts of the second study. Results of this study served to refine the framework of agricultural personnel management developed based on the first two studies. The most significant extension is an additional set of personnel management practices regarding the performance management function. Performance management describes the daily, informal interaction between managers and employees, including informal feedback, task-related communication, setting priorities, and dealing with problems. Although these practices are important in the day-to-day management processes, there has been little discussion about them in the literature. Also, working conditions were extended to include the organisational structure, and the social environment at work was established as another arena to be monitored and consciously managed. The resulting framework of agricultural personnel management includes eleven management functions: recruiting, selection, hiring immigrant employees, training, working conditions and organisational structure, social environment, performance management, discipline, performance appraisal, compensation, and labour law and regulation.

Managers' Personnel Management Competencies and Practices

In a recent study, Stup, Holden, and Hyde identified competencies in different management areas on the senior and the middle management level of dairy farms through group discussions and then surveyed different managers about their comfort level with respect to these competencies. While managers were generally confident about their competencies, senior managers were least confident about their personnel management competencies (4.95 on a 7-point Likert scale, 1=very low, 7=very high, n=41). Middle mangers ranked themselves second lowest in personnel management competencies (4.41, n=22) and lowest in community service and public relations (4.05, n=20).

Bitsch and Olynk (2007) developed a typology of required personnel management skills for successful management in animal agriculture based on ten focus groups with dairy and pork farmers and managers. The typology consists of five skill sets: motivator, housekeeper, model employee, counsellor, and change agent. This typology shows a number of commonalities with the Competing Values Framework, used in general management education (Faerman, Quinn, and Thompson), but also industry specific differences. The motivator with the ability to train and motivate others, and to provide constructive feedback and the housekeeper with the ability to control, to lead, and to discipline others build the core of agricultural personnel management skills and also likely other production enterprises. In addition, the ability and willingness to be a model employee plays a surprisingly large role in agriculture. The function of the counsellor, to support employees with their personal problems at work and beyond, was discussed less frequently by the research participants, but is necessary to prevent problems and to sustain employee productivity. The change agent initiates or implements innovations in the production process and was mentioned mostly by managers of larger farms. The authors point out that to be successful managers need to command a complete repertoire of skills including skills from each of the five types and not limit themselves to skills from only one type, for example, out of familiarity with certain behaviours (Hutt and Hutt).

The role and the functions of middle management are a field of agricultural personnel management with few studies, but increasing importance. Not only did the share and impact of hired labour increase with increasing farm sizes, and personnel management became more important, but supervisors and middle managers are also playing a larger role. Billikopf (2001) interviewed farm supervisors in California and found them to struggle with personnel management tasks. Bitsch and Yakura employed a case study approach to develop a grounded theory of agricultural middle management.

The participating middle managers described an unexpectedly large number of different personnel management practices. Bitsch and Yakura suggested that these practices can be clustered into two basic types: traditional practices and participative practices. Traditional practices include reprimanding employees, orienting and training employees, monitoring and controlling employees, and dealing with conflict. Participative practices include accommodating employees (*e.g.*, flexibility in schedules, task and team assignments), managing relationships with employees, providing information and goal setting, listening to employees, providing appreciation and feedback, rewarding employees (non monetary), modeling work behaviour, peer control, manager-induced team building, and training by coworkers.

Although this typology shows similarities with McGregor's Theory X and Y, Bitsch and Yakura underline a significant difference. For the participating middle managers, using traditional or participative practices was not correlated with individuals. Each manager used both traditional, as well as, participative practices. The authors suggest that management success corresponds rather with the number of practices individual managers command than with the type of practices they use more frequently. McGregor had assumed that participative managers would be more successful. Bitsch and Yakura pointed out that some managers did use few practices, whereas others were using the full breadth of the described practices. Given that day-to-day management consists of many different management situations, managers with a more complete repertoire are more likely to choose suitable practices.

Employees' Job Attitudes and Job Satisfaction

Job satisfaction is considered both a goal in itself, as well as, a means to reduce turnover and increase motivation and performance. Although meta-studies found a smaller relationship between job satisfaction and these correlates than expected, several studies of job satisfaction in agriculture have been conducted during the past 50 years. One of the more frequently applied models is the empirically grounded two-factor model by Herzberg et al. This model is particularly suited to structure the analysis of job attitudes and their context. Empirical evidence that indeed job satisfaction and job dissatisfaction are caused differently as predicted by the Herzberg et al. model is scant.

Independent of the theoretical models and the research methods several common results emerge from studies of job attitudes in agriculture. Porter pointed out that half of the dairy farm employees surveyed in New Hampshire saw appreciation of their work as the most important factor for their performance.

In addition, they mentioned open communication with their supervisor, good records, and control of the work situation; Porter concluded that financial incentives are less important. Adams, How, and Larson found financial incentives to be important for a satisfactory employer-employee relationships, but stressed the importance of consideration for workers as human beings, taking into account personal problems of workers and helping to find solutions, and getting the right fit of worker and job. Bitsch (1996) in a study of tree nursery apprentices in Germany found that a large majority did desire higher wages, but almost half also desired increased appreciation, more training, and more responsibility for their tasks. More training was also requested by Spanish speaking dairy farm employees surveyed by Maloney and Grusenmeyer in New York. Surveying New York dairy farms, Fogleman et al. found that employees were least satisfied with the factor managers had most control over, that is performance feedback.

Billikopf (2001) had found supervisors in all branches of agriculture to be mostly satisfied with their jobs. More detailed case studies with horticultural operations found for employees without supervisory responsibilities (Bitsch and Hogberg) and also for supervisors that the same factors seem to contribute to job satisfaction, as well as, to dissatisfaction, depending on their availability and characteristics. For both groups of employees, job security, achievement, technical competency of the superior, and personal relationships at the workplace were more likely to be perceived as positive. The work itself and organisational procedures and policies were perceived as ambiguous, contributing to both satisfaction and dissatisfaction.

Compensation was perceived rather negative, more negative by employees without supervisory responsibilities than by supervisors; the latter are likely to be higher paid and more likely to receive benefits. Employees without supervisory responsibility perceived their work/life balance more positive than supervisors; the latter are also less satisfied with their working conditions. Mainly, this was due to the fact that employees with supervisory responsibilities were expected to be available for work whenever required, whereas employees without supervisory responsibilities were given more flexibility. An earlier study in Germany, also had found that horticultural employees value flexible scheduling and benefit arrangements.

Relationships between Personnel Management Practices and Organisational Outcomes

Relationships between personnel management practices and various organisational outcomes, such as productivity (Rosenberg and Cowen), profit (Adams, How, and Larson), or competitiveness (Chacko, Walker, and Asar; Mugera and Bitsch) have often been assumed, but infrequently been empirically researched. Owners and managers of agricultural operations also testify to a relationship between personnel management practices and farm level outcomes (Bitsch et al.; Strochlic and Hamerschlag). The few studies attempting the empirical description and measurement of these relationships in production agriculture and agribusiness have found limited evidence.

Rosenberg and Cowen tested several personnel management practices' and management assumptions' impact on dairy farm productivity, including prevalence of Theory Y assumptions (McGregor), upward and downward responsibility diffusion, employee selection procedure, employee assessment criteria, and employee performance feedback, along with record use and herd size. In addition to record use, the authors found that Theory Y assumptions and the amount of feedback provided to employees impacted productivity. Feedback has also been found to be important in employees' job satisfaction. Although management assumptions are likely to guide organisational structure, personnel management practice choice, and managers' communication and interaction with employees, the study did not provide evidence of the relationship between assumptions and particular practices.

Stup, Hyde, and Holden analysed several personnel management practices of successful dairy farms in Pennsylvania, including milk quality incentives, performance reviews, employment of Spanish-speaking employees, use of standard operating procedures for milking, feeding, and reproduction tasks, continuing training, and use of job descriptions. Except for continuous training of employees, farm success did not differ significantly for farms using compared to farms not using these practices. While differences in definitions between Stup, Hyde, and Holden, and Rosenberg and Cowen and little overlap regarding the management practices researched, make it difficult to compare both studies, it should be noted that Stup, Hyde, and Holden did not find performance reviews to be significant.

Chacko, Wacker, and Asar compared perceptions of agribusiness managers with respect to the contributions of different technological and personnel management practices to their competitiveness. In general, managers ranked technological practices higher than personnel management practices. However, job security and measures of training and development were among the top ranked management practices. Job security has also been emphasized in job satisfaction studies. Training has been found to stand out in Stup, Hyde, and Holden and has also been emphasized in job attitude studies.

Based on managers' perception of particular technological and personnel management practices, Chacko, Wacker, and Asar also aggregated practices in a factor analysis and regressed these factors on perceived competitiveness. The regression analysis showed personnel management factors to contribute to a higher extent to different measures of competiveness than technological measures. The employee commitment factor (job security, sharing of profits and gains) stood out as contributing to most competitiveness measures.

Mugera and Bitsch used a resource based perspective to analyse whether personnel management practices and the personnel itself constitute a competitive advantage for dairy farms. The authors conducted case studies with dairy farms to analyse the integration of personnel management practices with each other (*e.g.*, practices regarding recruitment, selection, training, and compensation) and their outcomes (*e.g.*, voluntary turnover and termination). The case studies provided empirical examples of the applicability of the resource based theory and evidence of the use of personnel management practices as a competitive advantage. The authors emphasize that studies of isolated management practices may lead to misleading results, due to the importance of the integration of practices with each other. Therefore, they recommend an integrative approach to researching and changing personnel management functions.

Strochlic et al. surveyed 300 organic farms of various agricultural specialisations with respect to their personnel management practices and organisational outcomes. They found significant relationships between an overall labour conditions score and 5- and 10-year retention rates, several occupational safety related practices and person-days lost due to accidents and injuries. No relationship was found between the surveyed management practices and supervisory costs or access to sufficient labour.

Trends in Farm Productivity

Farm progress is normally regarded as a prerequisite of economic development. It is true that the economic development m the modem times has become associated with the industrialisation but it is generally accepted that industrialisation can follow only on the sound heels of farming. In our country farming is the largest sector of economic activity providing not only food and raw materials but also employment to a very large section of the population.

Being the dominant sector, farm output certainly has an effect on national output. Besides, our population is increasing at an alarming rate demanding similar progress in food production and other consumer goods which are farm based. This interaction between farm and non-farm sectors facilitates the growth of both. The demand for non-farm inputs of industrial origin stimulates industrial activity. If the industrial sector improves its growth rate the demand for the wages, goods and materials increases which in turn helps in the expansion of farm employment and earnings. Increase in farmearnings increases the demand for industrial goods and there could be a market developed for them. Thus,"development of farm sector depends on industrial development and vice versa.

Role of Agriculture in Indian Economy

The significant role of farming in the national economy can be described under the following heads.

Contribution to National Earnings

At the time of the first world war, farm contribution was about 2/3 road to the national earnings where there was not much industrial development. Its contribution has shown a declining trend since then. After the Independence, initiation of Five Year Plans stabilised the industrial infrastructure and thus the share of farming towards the national economy slowly decreased from 56.1 per cent (1950-51) to around 30 per cent (1993-94). The fact that farming still is an important contributor to the national earnings: it is often taken as an indicator of the economic development.

Normally, developed countries are less dependent on farming as compared to underdeveloped countries. For example, its contribution is only 3 per cent in the USA and U.K. which confirms that as the country progresses the dependence on farming decreases.

Largest Employment Providing Sector

Farming directly or indirectly has continued to be the main source of livelihood for the majority of the population. About 67 per cent -69 per cent of our country's working population is employed in different farm activities. It is only in developing and underdeveloped countries, that a higher percentage of the population is engaged in farming.

Contribution Towards Industrial Development

Indian farming is of prime importance in the industrial development as it supplies necessary raw materials for the concerned industries. Some depend on farming directly, like textile industries, sugar factories, etc.And some indirectly like breweries etc.

Experiences shows that the growth and diversification of farm production have helped to develop various types of industries and diversification of employment and a shortfall in farm growth for some time had an adverse effect on the industrial production affecting the prices causing imbalances in the economy. But the significance of farming in the industrial sector is decreasing as many industries have come up which are not dependent on farming.

Contribution to Foreign Exchange Resources

Farm products like tea, sugar, oil seeds, tobacco, spices etc. Occupy an important place in the country's export trade. The export statistics show that there have been in- crease from 1960s to till date. It was worth 7 crores during 1960-61 which has been increased to 958 crore rupees in 1993-94. Farming and allied products contributed 18.1 per cent in the total exports during 1993-94.

Role of Agriculture in Economic Planning

Importance of farming in the national economy is indicated by many facts. Its internal trading is the main support to our transport method. If there is a bumper production, its transport by road and railways makes business and if there is a depression it leads to loss in business. So, failure in farm front leads to the failure of economic planning.

Contribution to Capital Formation

Since it is the largest source of national earnings it is the primary source of savings and hence capital formation in the economy. Large investments in land development, construction of farm houses and irrigation facilities and other facilities like farm machinery and implements, warehouses, cold storages etc. Have been made after independence.

Both very- able and fixed capital in backward and progressive areas respectively can be built for necessary infrastructures. So, it is clear that farming has its own place in the national economy and its performance sets the pace of growth in the economy as a whole.

Harvesting Pattern in India

Harvest pattern has been defined as the proportion of area under different harvests at a particular period of time. A change in harvesting pattern means a change in the proportion of area under different harvests. It can be described in a number of ways but the most convenient method is to classify the farm production into two groups *i.e.*. Food grains and non-food grains, the details of which are-, as follows.

Harvesting Pattern of Food Grain Harvests

A large proportion of the area under food grains is occupied by cereals. The food grains occupied an area of 97.3 m ha. In 1950-51 has increased to 124.3 m ha. In 1970-71. From 1970-71 it increased to 126.7 and 127.8 million ha. in 1980-81 and 1990-91 respectively. But the figures in 1992- 93 shows a slight decrease to 123 m ha and 1993-94 figures showed a further decrease to 122.4 m ha. In 1993-94, the total area under cereals was 100 m ha. And under pulses was 22.4 m. Hectares. The area under each individual harvest of some food grains is given in table 1.

Table. Harvestping Pattern of Some Food Grains (Million Hectares).

Sl.No	Harvest	1970-71	1980-81	1990-91	1992-93	1993-94
1	Rice	37.6	40.1	42.7	41.8	42.0
2	Wheat	18.2	22.3	24.2	24.6	24.9
3	Jowar	17.4	15.8	14.4	13.0	12.9
4	Bajra	12.9	11.7	10.5	10.6	9.5
5	Maize	5.8	6.0	5.9	6.0	6.0
6	Other cereals	9.9	8.3	5.5	4.8	4.7
7	Gram	7.8	6.6	7.5	6.5	6.4
8	Tur	2.7	2.8	3.6	3.6	3.6
9	Other pulses	12.1	13.1	13.6	12.3	12.4

Rice is the major cereal harvest among food grains and showed a gradual increase in the area and so also the wheat. But coarse grains like jowar, bajra and maize showed a de- cline in the percentage of the area. If we study the area of cultivation of food grains and non-food grains, there was a gradual shift from non-food grains to food grains. Important reasons are: the prices of food grains have been rising quite fast and the farmers have started growing food harvests in the similar way they grow commercial harvests like cotton, oilseedharvestsugar cane etc. Secondly, the cultivation of food grains has become highly remunerative and productive under the impact of new technology.

Harvesting Pattern of Non-Food Grains

Among non-food grain harvests, oil seeds form an important group which also includes other harvests like cotton, jute, sugarcane, tobacco, tea, coffee, etc. The area has shown in- creasing and decreasing trends as given in Table below.

Table. Area Tender Some Important Non-Food Grains (Million Hectares).

Sl.No	Harvest	1970-71	1980-81	1990-91	1992-93	1993-94
1	Groundnut	7.3	6.8	8.3	8.2	8.4
2	Rapeseed and Mustard	3.3	4.1	5.8	6.2	6.3
3	Other oil seeds	6.0	6.7	10.0	10.9	12.1
4	Tobacco	0.4	0.4	0.4	0.4	N.A
5	Cotton	7.6	7.8	7.4	7.5	7.3
6	Jute	0.7	0.9	1.0	1.0	0.9
7	Sugarcane	2.6	2.7	3.7	3.6	3.4
8	Tea	0.4	0.4	0.4	0.4	0.4
9	Coffee	0.1	0.2	0.3	N.A	N.A

N. A. Available The table above shows that sometimes there is an increase in the area and sometimes there is a decrease in the area but overall there was not much change in the area of cultivation.

Factors Affecting Harvests Pattern in India

Harvesting pattern of any region depends upon physical characteristics as soil, climate, rainfall, etc. Apart from this, it depends on the nature and availability of irrigation facilities.

Besides, physical and scientific factors economic motivations are also important in determining the harvesting pattern. The prices influence the acreage under the harvests in two ways. One is that the variations in the interharvest price disparities led to shifts in acreage between the harvests. Another is that the maintenance of a stable level of prices for a harvest provides a better incentive to the producer to increase the opt put than what a very high level of price does, if there is no uncertainty of this level being maintained over a number of years.

The fixed procurement price of wheat and rice and other Government controls have induced farmers to shift the cultivation to cash harvests like sugar cane. Farmers also would choose the combination of harvests which would give them maximum earnings. Relative profitability per acre is the main consideration which influences the harvest pattern. Small farmers are first interested in producing food grains for their requirements and devote only a small relative acreage to cash harvests than large farmers. In fact, in recent years, it is the small farmer who have been increasing their sugarcane areas more than large farmers. Food Harvest Acts, Land use Acts, intensive schemes for paddy, cotton, oilseed" etc. All these bring sharply into focus the possibility that while each individual measure may push the harvest pattern in the direction intended to, but if the overall effect of all measures taken together with the entire harvest pattern is taken, it may not be in accordance with national requirements.

Productivity Trends in India

Before independence farm production rose only marginally as compared to the growth of population. With the introduction of economic planning in 1950-51, and with special emphasis on farm development, there was a steady increase in the area, productivity as well as in yield per hectare. Farm production in India *viz.* area, productivity per hectare and the total output is influenced by a large number of factors like rainfall, weather conditions, etc. The productivity figures of some important countries are given in Table below.

Table. Average Yield Per Hectare in Selected Countries.

Sl. No	Commodity	Country	1951-56	1961-65	1987-88
1	Rice (quintal/ha)				
		India	8	10	17
		China	17	18	35
		U.S.A	19	29	41
		Japan	26	33	40
2	Wheat (quintal/ha)				
		India	7	8	20
		China	9	9	30
		France	21	29	61
		Germany	28	33	68
3	Cotton (lint) (Kgs/ha)				
		India	90	120	202
		China	160	250	764
		USSR	-	700	787
		Mexico	336	640	1100

Before independence the productivity of food grains has shown a decline but with the introduction of planning in 1950- 51, there was a positive increase in the productivity. During the pre-green revolution period, rice recorded the most impressive growth rate in yield from merely 7 quintals per hectare in 1949-50 to 11 quintals by 1964-65.

The annual rate of growth was 2.1 per cent. The yield growth rate of wheat during the same period was increased from 6.6 q. to 9.1 quintals per hectare. Among non-food grains, cotton and sugarcane recorded modest growth rates.

During the post green revolution period, the most spectacular growth rate was recorded by wheat (3.2 per cent) and potato too recorded an impressive growth rate of 3.0 per cent per year.

Rice also registered a slow but steady rise of 2.1 per cent.Per year. In 1987-88, rice recorded an annual yield of 17.0 q/ha as against 41.0 q in USA and 40.0 q in Japan.

Table. Yield of Different Harvests (M. Tonnes).

Sl. No	Harvest	1993-94	19992-93	1990-91	1980-81	1970-71
	Food Grains	182.1	179.5	176.4	129.6	108.4
1	Cereals	169.0	166.6	162.1	119.0	96.6
	Rice	79.0	72.9	74.3	53.6	42.2
	Wheat	59.1	57.2	55.1	36.3	23.8
	Jowar	11.5	12.8	11.7	10.4	8.1
	Bajra	5.0	8.9	6.9	5.3	8.0
	Maize	9.5	10.0	9.0	7.0	7.5
	Others	4.9	4.8	5.1	6.4	7.0
2	Pulses	13.1	12,8	14.3	10.6	11.8
	Gram	4.9	4.4	5.4	4.3	5.2
	Tur	5.5	2.3	2.4	2.0	1.9
	Others	21.5	6.1	6.5	4.3	4.7
3	Oil seeds	21.5	20.1	18.6	9.4	9.6
	Groundnut (in shell)	7.8	8.6	7.5	5.0	6.1]
	Rape seed & Mustard	5.4	4.8	5.2	2.3	2.0
	Others	8.3	6.7	5.9	2.1	1.5
4	Tobacco	N.A	0.58	0.56	0.48	0.36
5	Cotton (lint)	10.7	11.4	9.8	7.0	4.8
6	Jute	7.4	7.5	7.9	6.5	4.9
7	Sugarcane	227.1	228.0	241.0	154.2	126.4
8	Tea	0.76	0.7	0.75	0.57	0.42
9	Coffee	0.21	0.16	0.17	0.12	0.11

Note: **Comprises ground nut, Reapeseed, mustard, seasomum, linseed, castor seed, niger seed, safflower, sunflower and soybean.

Even China recorded an average productivity of 35.0 q/ha. which was more than double of the average yield of India. Under the impact of the green revolution, the average annual yield per hectare in India was 20.0 q/ha in 1987-88 while it was 61.0 q/ha in France and 68.0 q/ ha.

In West Germany, Netherlands and UK have recorded average annual I yields of 81 and 69 q/ha respectively.

Table. Index Numbers of Farm Production and Productivity.

Sl. No	Production	1993-94	1992-93	1990-91	1980-81	1970-71
		Triennium ending 1981-82 = 100				
1	All commodities	154.8	151.5	148.4	102.1	85.9
	Food grains	148.3	144.3	143.7	104.9	87.9
	Non food grains	165.8	163.6	156.3	97.4	82.6
2	Selected harvests					
	Rice	158.8	146.5	149.4	107.8	84.4
	Wheat	168.0	162.5	56.6	103.2	67.7
	Gram	119.2	107.4	180.2	105.4	126.3
	Groundnut	129.4	142.8	125.3	83.4	101.8
	Cotton	142.4	151.6	130.9	93.2	63.4
	Jute	121.6	138.3	122.6	100.8	76.5
	Sugarcane	145.3	145.9	154.3	98.8	81.2
	Tea	125.5	125.5	132.3	101.6	74.7
	Coffee	116.2	116.2	122.3	85.1	79.1
	Tobacco	118.5	124.2	115.8	100.2	75.5
3	Productivity					
	All commodities	139.7	137.0	133.8	102.9	92.6
	Food grains	145.5	142.0	137.8	105.9	93.2
	Non food grains	131.7	130.0	128.0	99.2	91.4
4	Selected harvests					
	Rice	151.4	140.6	140.2	107.7	90.2
	Wheat	149.6	146.7	143.8	102.8	82.4
	Gram	132.2	118.8	123.6	114.3	115.3
	Groundnut	110.0	124.6	107.4	87.5	99.0
	Cotton	155.4	160.9	140.8	95.4	66.7
	Jute	137.7	137.0	136.6	92.9	88.6
	Sugarcane	121.3	115.5	118.3	104.6	88.8
	Tea	113.6	113.6	119.9	101.1	80.2
	Coffee	116.4	116.4	110.2	184.6	122.0
	Tobacco	134.9	130.7	124.0	97.7	74.4

Between 1961 and 1988, the yield of wheat per hectare had gone up by only 150 per cent ha India as against 233 per cent in China. In case of cash harvest cotton, our country's per hectare yield was only 202 kgs. as against 764 kgs. in China, 787 kgs in USSR and 1100 kgs. in Mexico (1987-88). Likewise, maize recorded only 14.0 q/ha in 1987-88 as compared to 37.0 q/ha in China and 70.0 q/ha in the USA. So, the basic conclusion is though there is a rise in productivity, the average yield per hectare is below the world average in all harvests.

Table. Yield of Selected Harvests.

Sl. No		1993-94	1992-93 (Kgs/ha)	1990-91	1980-81	1970-71
1	Food grains	1487	1457	1380	1023	872
	Kharif	1308	1302	1231	933	837
	Rabi	1781	1725	1635	1195	942
	Cereals	1690	1654	1571	1142	949
	Kharif	1456	1440	1357	1015	892
	Rabi	2116	2068	2010	1434	1093
	Pulses	584	573	578	473	524
	Kharif	476	495	471	361	410
	Rabi	696	654	672	571	607
	Rice	1879	1744	1740	1336	1123
	Kharif	1797	1677	1670	1303	1100
	Rabi	2814	2653	2671	2071	1625
	Wheat	273	2327	2281	1680	1307
	Jowar	894	982	814	660	466
	Kharif	1084	1030	969	937	533
	Rabi	672	632	582	520	354
	Maize	1583	1676	1518	1159	1279
	Bajra	576	836	658	458	622
	Gram	761	684	412	657	663
	Oilseeds	801	797	771	532	579
	Kharif	753	804	698	492	649
	Rabi	875	780	872	588	449
	Groundnut	926	1049	904	736	834
	Sugarcane (tonnes)	67	64	65	58	48
	Cotton	248	457	225	152	106
	Jute	1907	1857	1833	1245	1186

From the table, it is observed that during the period between 1980-88, other countries have shown rising in productivity for example: between 1965 -88, China has shown an increase of 94 per cent in rice yield (per hectare) whereas India's was only 50 per cent. As it was already mentioned wheat per hectare yield was only 180 per cent as against 233 percent. In cotton harvest, there was an increase of 250 per cent in China as compared to 63 per cent in India. The farm production trend in different harvests in

India is given in Table. Cotton in million bales of 170 kg. Each. Jute in million bales of 180 kg. Each. The food grains production in 1994-95 was 189.8 m. tones and it was 191.8 million tones in 1995-96. The index numbers of farm production and productivity is given in Table below.

The yield of some selected harvests of our country can also be seen in table below. Even though there has been substantial growth in farm production, this has not been smooth, instead, there have been continuous fluctuations in harvest output from year to year. As already explained India has low farm productivity. There are several causes for low productivity.

Causes of Low Productivity

The farm productivity in India *i.e.* average yield per hectare is among the lowest in the world. There has been some progress in all the sectors in recent years particularly during the plan period. But there is not much change in the farmsector, especially when we consider the condition of the farmers. The analysis of the factors that are responsible for low productivity may be helpful in improving the situation. Some of the main factors are briefly described here.

Overcrowding in Agriculture

The main problem in the farm sector is too many people are directly dependent on it though all depending on it indirectly for their existence. The national increase in population could not be absorbed in industries and people engaged in other sectors like handiworks etc. has also adopted farming.

Consequently, the pressure of population increased the pressure on land. Also it has led to sub-division and fragmentation of holdings thus there was a decline in the area of land available for cultivation per capita, disguised unemployment in farming and low marginal productivity of labour. The pressure of increased population was so high that between 1901 and 1981, the cultivable area per cultivator has declined from 0.43 hectares to 0.23 hectares. So it is essential to check the growth of the rural population and decrease the pressure on land.

Discouraging Rural Atmosphere

In general, Indian farmers are mostly illiterate ignorant, superstitious and conservative and they feel satisfied with their primitive method of cultivation. Only very few farmers are quick in following modern technology exposed to them; but the vast majority of farmers is not motivated to learn and try new ways. But this atmosphere is slowly changing and unless this is changed there will be no progress in the farm sector.

Inadequate Non-farm Services

In the farm sector, there were inadequate non-farm services like finance, marketing, etc. Till the recent past, fanners had to depend on the village money lenders paying high rates of interest and there was a risk of losing his land and become landless labourer. Other sources of finance did exist such as cooperatives and the Government but they were almost insignificant. Similarly, the farmers could not secure storage facilities in towns and sometimes they were cheated by wholesalers and commission agents.

Size of Holdings

The average size of holdings in India is very low and they are fragmented and small. Since they are small, scientific cultivation techniques cannot be adopted; Small sized holdings lead to a great waste of time, labour and cattle power, difficulty in proper utilisation of irrigation facilities and consequent litigation among farmers etc. So the small holdings are one of the causes for poor farm yield.

Pattern of Land Tenure

The zamindarimethod which is abolished now was affecting the tenant farmer. Now, tenancy legislations have been established to protect the tenant farmers. But he has to pay high rent and he has no security and he can be turned out of his land at any time by the landlord. Under these circumstances, it is impossible to expect the tiller to increase farm productivity.

Poor Techniques of Production

Since our farmers are tradition-bound and poor they have not aborted the techniques which are so widely adopted in the countries of the West and Japan. Only in recent years and that too to a limited extent, the farmers have started adopting improved implements like steel ploughs, sugarcane. Crushers, small pumping sets, water lifts, hose, seed drills etc. Increase in production is only possible if proper and adequate manures are used. To revitalise the fertility and to utilise fallow lands, the use of various kinds of manures is essential. Besides, use of FYM and chemical fertilizers is inadequate. About 10-20 per cent increase in productivity can be brought through the use of better seeds. Our farmers have no access to better seeds and also lack of storage facilities for the preservation of the same for sowing.

Inadequate Irrigation Facilities

Indian farming is mostly dependent on rainfall and very few farmers avail the facility of artificial irrigation. Though the more area was brought under irrigation, still there is a great scope for improving the irrigation facilities.

Besides these enumerated causes the following are also the causes for poor productivity in farming.

Law of

According to the law of all the children of a property holder demand an equal share resulting in endless fragmentation of the cultivated lands. If a man has 3 to 4 plots of varying soil properties and productivity all the children demand portions of each of the plots in the name of equality in sharing. This practice of equal division has gone to such an extent that we come across people in some areas owning a number of plots in different places. Some of these plots are as big as an office table or a small living room. Obviously it is impossible to do any worthwhile cultivation on these plots situated in different places.

Poor Educational Standard

Though many farmers are considered literate after implementation of the National Adult Literacy Mission most of them are merely able to sign their names only. They are not able to write their own ideas nor are they able to do the basic arithmatics. They also do not have any basic general knowledge of biology, chemistry or physics which is necessary for understanding various agronomic practices in the farm. Ignorance breeds superstition and superstition inhibits growth not only personal but also economy. Even the so called educated in our country do not have basics of scientific farming. Rural youth when they fail to get any other job they turn to farming. But not knowing the basic they become mere manual labourers.

Non-viable Farm Holdings

For a farm to be economically viable, technically feasible and environmentally sound it should have a minimum size ego 10-15 acres or above. Only then a family can produce enough net profit and make a decent living on it which implies that it maintains a satisfactory standard of productivity.

Inadequate Prices

The prices of farm products are fixed by the government officials whereas that of the non-farm products is fixed by the producers. If we cost account all the expenses the production per unit farm product is several times higher than the price fixed by the government. That means a farmer is cultivating this land always in the loss. If a man is loosing in his business how can he improve his business. So also who cultivates in loss can not improve his productivity.

Caste Factor

Cultivation is mostly done by low caste people in India. People engaged in manual work in field are looked down upon. They are there because they have no other way of living. Under such socially oppressive situation productivity cannot be improved.

Human Energy Requirement

To carry out the usual farming operation one requires to spend high amounts of his physical energy. Long hours of hard and strenuous work combined with poor working conditions and low level of food energy availability and less time for reoccupation a farmer gets constantly worn out. As a result he will not be able to improve the farm or his labour productivity.

Lack of Infra-structural Facilities

Most of the farmlands are situated in the areas where there are no road and transportation facilities nor any communication facilities. Hence unable to transport his products in time nor he can buy his requirements.

Lack of Personal Security

The farmer (the cultivator) has probably the least social security. He has no paid holidays no allowances (medical, child, educational, festival, travel etc.)

He has no retirement benefits nor his age of retirement fixed. He retires only at his death. In this situation how a farmer makes efforts to increase the productivity.

Multifarious Responsibilities

The farmer has too many roles and responsibilities. He is the owner! Tenant! cultivator, labourer, purchaser, store keeper, transporter, seller, buyer, accountant, manager, investor of investor, agronomist, pathologist, accountant, manager, pest controller, soil and water conservationist, seed collector, seed processor, storer, mechanic, risk taken etc. Humanly speaking it is not possible for anyone to perform all these roles and responsibilities. Every sphere of organised human activities, we find that each person is doing only the specialised and specified jobs for a specified period of time. Whereas the farmers have to bear the burden of many responsibilities at the same time. Nobody seems to realise this; the pitiable thing is that even the farmer himself is not realised. Under multifarious roles and responsibilities it is not possible for anyone to think of increasing the productivity of harvests.

General Suggestions for Progress.

In order to seek a change in the farmers, it is the duty of the Government, other related sectors and organisations to make available all the resources accessible to the farmers. For implementing the necessary programmes for a better harvesting pattern, suggestions has to be given tobeneficial harvest rotation suitable to that area. Cooperation from the farm officers is expected which will help the farmers and establishing a Farm Mechanisation Corporation there by an average farmer who cannot manage with hired labour would be benefited. Government should give the greatest importance to the promotion of transport, marketing facilities and consolidation of holdings.

Progress in storage facilities, providing tenant security, implementing the recommended projects in rural areas for improving the irrigation facilities for supplementing better quality seeds and for increasing awareness among the people about hybrids, varieties, disease resistant varieties, drought resistant varieties, etc. are the other major aspects to be concentrated upon. There is no scope for increasing the area of cultivation in the future but through multiple harvesting, relying on irrigation facilities, high yielding varieties, etc. We can raise the farm productivity.

Conclusion

To increase the growth rate it is important to change the harvesting pattern which can be achieved through appropriate changes in economic motive. A conservative farmer does accept the logic for a change wherever he has shown a better harvesting pattern. It is possible only when he feels supported and secured by the government economically and Socially. He should get adequate remuneration for his labour.

Farm Marketing in India

In India Farming was practiced formerly on a subsistence basis; the villages were self sufficient, people exchanged their goods, and services within the village on a barter basis. With the development of the means of transport and storage facilities, farming has become commercial in character, the farmer grows those harvests that fetch a better price. Marketing of farm produce is considered as an integral part of farming, since an agriculturist is encouraged to make more investment and to increase production. Thus there is an increasing awareness that it is not enough to produce a harvest or animal product; it must be marketed as well.

Farm marketing involves in its simplest form the buying and selling of farm produce. This definition of farm marketing may be accepted in olden days, when the village economy was more or less self-sufficient, when the marketing of farm produce presented no difficulty, as the farmer sold his produce directly to the consumer on a cash or barter basis. But, in modem times, marketing of farm produce is different from that of olden days. In modem marketing, farmproduce has to undergo a series of transfers or exchanges from one hand to another before it finally reaches the consumer. The National Commission on Farming, defined farm marketing as a process which starts with a decision to produce a saleable farm commodity and it involves all aspects of the market structure of the method, both functional and institutional, based on technical and economic considerations and includes pre and post- harvest operations, assembling, grading, storage, transportation and distribution. The Indian council of Farm Research defined involvement of three important functions, namely (a) assembling (concentration) (b) preparation for consumption (process) and (c) distribution.

Importance and Objectives of Agriculture Marketing

The farmer has realised the importance of adopting new techniques of production and is making efforts for more earnings and higher standards of living. As a consequence, the harvesting pattern is no longer dictated by what he needs for his own personal consumption but what is responsive to the market in terms of prices received by him. While the trade is very organised the farmers are not Farmer is not conversant with the complexities of the marketing method which is becoming more and more complicated. The cultivator is handicapped by several disabilities as a seller. He sells his produce at an unfavorable place, time and price.

The objectives of an efficient marketing method are:

1. To enable the primary producers to get the best possible returns,
2. To provide facilities for lifting all produce, the farmers are willing, to sell at an incentive price,
3. To reduce the price difference between the primary producer and the ultimate consumer, and
4. To make available all products of farm origin to consumers at reasonable prices without impairing on the quality of the produce.

Facilities Needed for Farm Marketing

In order to have the best advantage in marketing of his farm produce the farmer should enjoy certain basic facilities.

1. He should have proper facilities for storing his goods.
2. He should have holding capacity, in the sense, that he should be able to wait for times when he could get better prices for his

produce and not dispose of his stocks immediately after the harvest when the prices are very low.

3. He should have adequate and cheap transport facilities which could enable him to take his surplus produce to the Mandi rather than dispose it in the village itself to the village money-lender-cum-merchant at low prices.
4. He should have clear information regarding the market conditions as well as about the ruling prices, otherwise may be cheated. There should be organised and regulated markets where the farmer will not be cheated by the "Dallas" and "arhatiyas".
5. The number of intermediaries should be as small as possible, so that the middleman's profits are reduced. This increases! the returns to the farmers.

Inadequacies of Present Marketing Method

Indian method of farm marketing suffers from a number of defects. As a consequence, the Indian farmer is deprived 'of a fair price for his produce. The main defects of the farm marketing method are discussed here.

Improper Warehouses

There is an absence of proper warehousing facilities in the villages. Therefore, the farmer is compelled to store his products in pits, mud-vessels, "Kutcha" storehouses, etc. These unscientific methods of storing lead in considerable wastage. Approximately 1.5 per cent of the produce gets rotten and becomes unfit for human consumption.

Due to this reason supply in the village market increases substantially and the farmers are not able to get a fair price for their produce. The setting up of Central Warehousing Corporation and State Warehousing Corporation has improved the situation to some extent

Lack of Grading and Standardisation

Different varieties of farm produce are not graded properly. The practice usually prevalent is the one known as "Dara" sales wherein heap of all qualities of the produceis sold in one common lot Thus the farmer producing better qualities is not assured of a better price. Hence there is no incentive to use better seeds and produce better varieties.

Inadequate Transport Facilities

Transport facilities are highly inadequate in India. Only a small number of villages are joined by railways and Pucca roads to mandies.

Produce has to be carried on slow moving transport vehicles like bullock carts. Obviously such means of transport cannot be used to carry produce to far-off places and the farmer has to dump his produce in nearby markets even if the price obtained in these markets is considerably low. This is even more true with perishable commodities.

Presence of a Large Number of Middlemen

The chain of middlemen in the farm marketing is so large that the share of farmers is reduced substantially. For instance, a study of D.D. Sidhan revealed, that farmers obtain only about 53 per cent of the price of rice, 31 per cent being the share of middlemen (the remaining 16 per cent being the marketing cost). In the case of vegetables and fruits the share was even less, 39 per cent in the former case and 34 per cent in the latter.

The share of middle- men in the case of vegetables were 29.5 per cent and in the case of fruits was 46.5 per cent. Some of the intermediaries in the farm marketing method are -village traders, Kutcha arhatiyas, Puccaarhatiyas, brokers, wholesalers, retailers, moneylenders, etc.

Malpractices in Unregulated Markets

Even now the number of unregulated markets in the country is substantially large. Arhatiyas and brokers, taking advantage of the ignorance, and illiteracy of the farmers, use unfair means to cheat them. The farmers are required to pay arhat (pledging charge) to the arhatiyas, "tulaii" (weight charge) for weighing the produce, "palledari" to unload the bullock-carts and for doing other miscellaneous types of allied works, "garda" for impurities in the produce, and a number of other undefined and unspecified charges. Another malpractice in the mandies relates to the use of wrong weights and measures in the regulated markets. Wrong weights continue to be used in some unregulated markets with the object of cheating the farmers.

Inadequate Market Information

It is often not possible for the farmers obtain information on exact market prices in different markets. So, they accept, whatever price the traders offer to them.

With a view to tackle this problem the government is using the radio and television media to broadcast market prices regularly. The newspapers also keep the farmers posted with the latest changes in prices. However the price quotations are sometimes not reliable and sometimes have a great time-lag. The trader generally offers less than the price quoted by the government news media.

Inadequate Credit Facilities

Indian farmer, being poor, tries to sell off the produce immediately after the harvest is harvested though prices at that time are very low. The safeguard of the farmer from such "forced sales" is to provide him credit so that he can wait for better times and better prices. Since such credit facilities are not available, the farmers are forced to take loans from money lenders, while agreeing to pledge their produce to them at less than market prices. The cooperative marketing societies generally cater to the needs of the large farmers and the small farmers are left at the mercy of the money lenders.

Thus it is not possible to view the present farm marketing method in India in isolation of (and separated from) the land relations. The regulation of markets broadcasting of prices by All India Radio, progress in transport method, etc., Have undoubtedly benefited the capitalist farmers, and they are now in a better position to obtain favorable prices for their "market produce" but the above mentioned changes have not benefited the small and marginal farmers to any great extent.

Characteristics of FarmProduce

Farm products differ in nature and contents from industrial goods in the following respects.

- Farm products tend to be bulky and their weight and volume is great for their value in comparison with many industrial goods.
- The demand for storage and transport facilities is more heavy, and more specialised in the case of farm products than in the case of manufactured commodities.
- Farm commodities are comparatively more perishable than industrial goods. Although some harvests such as rice and paddy retain their quality for a long time, most of the farm products are perishable and cannot remain long on the way to the final consumer without suffering loss and deterioration in quality.
- There are certain farm products such as mangoes and grapes which are available only in their seasons but this condition of seasonal availability is not found in the case of industrial goods.
- Farm produce is to be found scattered over a vast geographical area and as such its collection poses a serious problem. But such is not conditioned in the case of industrial goods.
- There are various kinds and varieties in farm produce and so it is difficult to grade them.
- The farmers especially in countries like India has low holding-

back. Therefore he has to sell his produce immediately after the harvest at whatever price he can fetch because of his pressing needs.

- Finally, both demand and supply of farm products are inelastic. A bumper harvest, without any minimum guaranteed support price from the government may spell disaster for the farmer. Similarly the farmer may not really be in a position to take advantage of shortages or deficit harvest. These benefits may pass on only to the middleman.

Methods of Sale and Marketing Agencies

The marketing of farm produce is generally transacted in one of the following ways.

Undercover or the Hatta Method

Under this method, the sale is effected by twisting or clasping the fingers of the seller's agent under cover of a cloth. The cultivator is not taken into confidence until the final bid is cleared.

Open Auction Method

Under this method the agent invites bids for the produce and to the highest bidder the produce is sold.

Dara Method

Another related method is to keep the heaps of grains of different quantities and sell them at fiat rates without indulging in weightment etc.

Moghum Sale

Under this method, the sale is based on the verbal understanding between buyers and sellers and without mentioning the rate as it is understood that the buyers will pay the prevailing rate.

Private Agreement

The seller may invite offers for his produce and may sell to one who might have offered the highest price for the produce.

Government Purchase

The government agencies lay down fixed prices for different qualities of farming commodities. The sale is effected after a gradual process for graduation and proper weightment. This practice is also followed in co-operative and regulated markets.

Marketing Agencies

The various agencies engaged in the marketing of farm produce can be classified into two categories, *viz.*, (i) Government and quasi private agencies like the co-operative societies and (ii) private agencies. A chain of middlemen may be found operating both in Government and private agencies.

Most important among these are as follows:

a. The merchant is the most usual purchaser of the product, he deals in his individual capacity.
b. Itinerant Beoparis (merchants) visit different villages, Collect the produce, and take to the nearest market.
c. Take the weighing men from the villages to the dealers in town.
d. Agents are concerned with the assembling and distribution of farm produce.

Farm Marketing in India

The existing methods of farm marketing in India are as briefly described here.

Sale to Financiers and Traders

A considerable part of the total produce is sold by the farmers to the village traders and financiers. According to an estimate 85 per cent of wheat, 75 per cent of oil seeds in U.P., 90 per cent of jute in West Bengal and 60 per cent of wheat, 70 per cent of oil seeds and 35 per cent of cotton in Punjab are sold by the farmers in the villages themselves. Often the money lenders act as a commission agent of the wholesale trade.

Hats and Shanties

Hats are village markets often held once or twice a week, while shanties are also village markets held at longer intervals or on special occasions. The agents of the wholesale merchants, operating in different mandies also visit these markets. The area covered by a "hat" usually varies from 5 to 10 miles. Most of "hats" are very poorly equipped, are uncovered and lack storage, drainage, and other facilities. It is important to observe that only small and marginal farmers sell their produce in such markets. The big farmers with large surplus go to the larger wholesale markets.

Mandies or Wholesale Markets

One wholesale market often serves a number of villages and is generally located in a city. In such mandies, the business is carried on by arhatiyas. The farmers sell their produce to these arhatiyas with the help of brokers, who are generally the agents of arhatiyas. Because of the malpractices of these middlemen, problems of transporting the produce from villages to mandies, the small and marginal farmers are hesitant about coming to these mendies.

The arhatiyas of these mendies sell off the produce to the retail merchants. However, paddy, cotton and oilseeds are sold off to the mills for processing. The marketing method for sugarcane is different. The farmers sell their produce directly to the sugar mills.

Co-operative Marketing

To improve the efficiency of the farmmarket and to save the farmers from the injustice and malpractices of middlemen, emphasis has been laid on the development of cooperative marketing societies. Such societies are formed by farmers to take advantage of collective bargaining. A marketing society collects surplus from it's members and sell it in the Mandi collectively. This improves the bargaining power of the members and they are able to obtain a better price for the produce. In addition to the sale of produce, these societies also serve the members in a number of other ways.

Progress of Farm Marketing Method

Government of India has adopted a number of measures to improve farm marketing, the important ones being - establishment of regulated markets, construction of warehouses, provision for grading, and standardisation of produce, standardisation of weight and measures, daily broadcasting of market prices of farmharvests on All India Radio, progress of transport facilities, etc.

Marketing Surveys

In the first place the government has undertaken marketing surveys of various goods and has published these surveys. These surveys have brought out the various problems connected with the marketing of goods and have made suggestions for their removal.

Grading and Standardisation

The government has done much to grade and standardise many farm goods. Under the Farm Produce (Grading and Marketing) Act the Government has set up grading stations for commodities like ghee, flour, eggs, etc. The graded goods are stamped with the seal of the Farm Marketing Department -Agmark The «Agmark" goods have a wider market and command better prices.

A Central Quality Control Laboratory has been set up in Nagpur and eight other regional laboratories in different parts of the country with the purpose of testing the quality and quality of farm products applying for the Government's "Agmark" have been created The Government is further streamlining quality control enforcement and inspection and progress in grading.

Organisation of Regulated Markets

Regulated markets have been organised with a view to protect the farmers from the malpractices of sellers and brokers. The management of such markets is done by a market committee which has nominees of the State Government, local bodies, arhatiyas, brokers and farmers. Thus all interests are represented on the committee. These committees are appointed by the Government for a specified period of time.

Important functions performed by the committees can be summarised as follows:

a. Fixation of charges for weighing, brokerages etc.,
b. Prevention of unauthorised deductions, underhand dealings, and wrong practices by the artist,
c. Enforcing the use of standardised weights,
d. Providing up to date and reliable market information to the farmers, and
e. Settling of disputes between the parties arising out of market operations.

The method of regulated markets has been found to be very useful in removing fraudulent practices followed by brokers and commission agents and in standardising market practices. The committee is responsible for the licensing of brokers and weightmen. It is nested with powers to punish anyone who is found guilty of dishonest and fraudulent practices. 1t is the policy of the government to convert all markets in the country into the regulated type.

Regulated markets aim at the development of the marketing structure to have the following:

1. Ensure remunerative price to the producer of farm commodities,
2. Reduce non-functional margins of the traders and commission agents, and
3. Narrow down the price spread between the producer and the consumer.

To achieve these objectives, the government would go in for comprehensive and rapid expansion of regulated marketing methods. The success achieved in states like Punjab and Haryana, where regulated markets have been established in major producing areas with linked up satellite markets in the rural growth centers would be aimed at, in other areas where intensive production is taken up. The regulating marketing method has also proved a good source of generating earnings for the marketing boards and for use in rural infrastructure. The regulated market complex will also include facilities for grading and for monitoring of prices.

The development of regulated markets is proposed especially in areas where commercial harvests like cotton, jute, tobacco and important non-traditional harvests are produced and sold in weekly markets and hats. Co-operative marketing and distribution and banking will also be linked to the regulated markets. These markets will cover all the major harvests. Separate market yards are proposed for livestock, fish, fruits and vegetables. There are now over than 6,050 regulated markets with the establishment of these regulated markets. The malpractices in mandies having disappeared and the market charges have been rationalised. As much as 70 per cent of farm produce is now sold in regulated markets. In this connection, the steps taken to standardise the weight and measures in the country should be mentioned. The government has successfully replaced the different methods of weights and measures prevalent in the country with the metric method.

Provision of Warehousing Facilities

To prevent distress sale by the farmers, particularly the small and marginal farmers, due to prevailing low prices, rural go downs have been set up. The government has done much to provide warehousing in towns and villages.

The Central Warehousing Corporation was set up in 1957 with the purpose of constructing and running go down and warehouses for the storage of farm produce. The states have set-up the State Warehousing Corporations with the same purpose. At present the Food Corporation is constructing its own network of go downs in different parts of the country. The total storage capacity in the country was 27 million tonnes at the end of the sixth plan.

Dissemination of Market Information

The government has been giving attention to the broadcasting of market information to the farmers. Since most villages have radio sets, these broadcasts are actually heard from farmers. The newspapers also publish farm prices either daily or weekly accompanied by a short review of trends.

Directorate of Marketing and Inspection

The directorate was set up by the Government of India to co-ordinate the farm marketing of various agencies and to advise the Central and State Governments on the problems of farm marketing.

Activities of this directorate include the following:

- Promotion of grading and standardisation of farm and allied commodities;
- Statutory regulation of markets and market practices;
- Training of personnel;
- Market extension;
- Market research, survey and planning
- Administration of old storage order, 1980 and the Meat Food Products order, 1973.

The directorate has so far formulated grade specification for 142 farm commodities. It enforces compulsory quality control before export on as many as 41 farm commodities. It is extending financial assistance to selected regulated markets for providing grading facilities for important commodities like tobacco, jute, cotton, groundnut and cashew nut at the producer level. An allied task is the one related to marketing research and survey. This should aim at determination of best handling methods to produce to minimize losses, damage and costs, improved methods of wholesaling and retailing and planning for new marketing facilities at appropriate centers. With this aim in view, the Directorate is currently implementing two schemes.

- Market research and planning.
- Market planning and design.

Under the former scheme, the Directorate has been carrying out countrywide marketing surveys on livestock and important farm and horticultural commodities to identify and study the problems of farm marketing. Under the latter scheme, the Directctorate has set up a Marketing Planning and Design Centre at Faridabad and a training center and Workshop in Nagpur to study the packaging grading and marketing of selected fruits and vegetable and also advise the authorities on the designing of fruits and vegetable markets.

Government Purchases and Fixation of Support Prices

In addition to the measures mentioned above, the Government also announces minimum support price for various farm commodities from time to time in a bid to ensure fair returns to the farmers. These prices are fixed in accordance with the recommendations of the Farm, Price Commission.

If the prices start falling below the declared level (say, as a result of a glut in the market), the Government agencies like the Food Corporation of India intervene in the market to make a direct purchase from the farmers at the support prices. These purchases are sold off by the Government at reasonable price through the public distribution method.

Cooperative Marketing

Though the above measures have improved the method of farm marketing to some extent, a major part of the benefits has been derived by large farmers, who have an adequate marketable surplus. However, the small and marginal farmers continue to sell a major part of their produce to financiers to meet their credit needs and these financiers offer them very low prices. Therefore it is essential to form cooperatives of the small and marginal farmers to enable them to obtain fair prices for their produce. The advantages that co-operative marketing can confer on the farmer are multifarious, some of which are listed below.

Increases Bargaining Strength of the Fanners

Many of the defects of the present farm marketing method arise because often one ignorant and illiterate farmer (as an individual) has to face well-organised mass of clever intermediaries. If the farmers join hands and for a cooperative, naturally they will be less prone to injustice and malpractices. Instead of marketing their produce separately, they will market it together through one agency.

Direct Deals with Female Buyers

In cases, the co-operatives can altogether skip the intermediaries and enter into direct relations with the final buyers. This practice will eliminate exploiters and ensure fair prices to both the producers and the consumers.

Provision of Credit

The marketing co-operative societies provide credit to the farmers to save them from the necessity of selling their produce immediately after harvesting. This ensures better returns to the farmers.

Easier and Cheaper Transport

Bulk transport of farm produce by the societies is often easier and cheaper. Sometimes the societies have their own means of transport. This further reduces cost and botheration of transporting produce to the market.

Storage Facilities

The cooperative marketing societies generally have storage facilities. Thus the farmers can wait for better prices. Also there is no danger to their harvest yield from rains, rodents and thefts.

Grading and Standardisation

This task can be done more easily for a cooperative agency than for an individual farmer. For this purpose, they can seek assistance from the government or can even evolve their own grading arrangements.

Market Intelligence

The co-operatives can arrange to obtain data on market prices, demand and supply and other related information from the markets on a regular basis and can plan their activities accordingly.

Influencing Marketing Prices

While previously in the market prices were determined by the intermediaries and merchants and the helpless farmers were mere spectators force to accept, whatever was offered to them, the co-operative societies have changed the entire complexion of the game. Wherever strong marketing cooperativeis operative, they have bargained for and have achieved, better prices for their farm produce.

Provision of Inputs and Consumer Goods

The cooperative marketing societies can easily arrange for bulk purchase of farm inputs, like seeds, manure fertilizers etc. And consumer goods at relatively lower price and can then distribute them to the members.

Processing of Farmproduces

The co-operative societies can undertake processing activities like crushing seeds, ginning 'and pressing of cotton, etc.

In addition to all these advantages, the co-operative marketing method can arouse the spirit of self-confidence and collective action in the farmers without which the programme of farm development, howsoever well conceived and implemented, holds no promise of success.

Warehousing in India

Warehousing facilities are necessary to prevent the loss arising out of defective storage and also to equip the farmers with a convenient instrument of credit. Both the Farm Finance Subcommittee (1945) and the Rural Banking Enquiry Committee (1950) emphasized the importance of warehousing as a method of promoting rural banking and finance in India.

All India Rural Credit Survey Committee (1954) recommended a three tire method of warehousing: at the national level, state and district level, village and rural level.

At present there are three main agencies in the public sector which are engaged in building large scale storage/warehousing capacity. They are: the Food Corporation of India (FCI), Central Warehousing Corporation

(CWC), and State Warehousing Corporation (SWC). FCI provides storage capacity for food grains. It has its own go downs and it also hires storage capacity from other sources such as CWC, SWC's, State Governments and private parties. In 1960-61, there were only 40 general warehouses in the countries with a total capacity of less than 0.1 million tonnes. By the end of 1988-89, the three public sector units have a storage capacity of nearly 32 million tonnes.

Besides, public sector agencies, co-operatives have also constructed warehouses in rural areas for storage of their members' produce, for stocking of fertilizers and other inputs and consumer articles. To avoid unfair competition with the go downs of the cooperative marketing societies, the state warehousing corporations do not open warehouses at any place below the sub divisional level. By 1987-88, a total storage capacity of over 10 million tonnes in the co-operative sector was available.

Kinds of Warehouses

There are broadly speaking four kinds of warehouses.

They are:

a. Private warehouses which are usually maintained by joint-stock companies, firms and individuals;
b. Duty-paid public warehouses which are maintained by dock authorities or port trust authorities at the port
c. Bonded warehouses which are maintained either by dock authorities or by the Government and
d. Licensed warehouses which are private warehouses run by co-operative societies or by private agencies, after obtaining a license from the Government.

Benefits of the Warehouses

The following are some of the benefits of the warehouses:

- It gives withholding power to the agriculturist to tide over difficulties and helps them to secure better prices for their produce.
- It gives purchasing power to traders.
- It tends to cushion the price fluctuation and stabilise prices as it equates supply to demand.
- It facilitates future trading.
- It plays a very important role in implementing the farm price policy of the Government
- It obviates the need for unnecessary cross-transport.

- Huge wastages which occur owing to improper storage of farm produce will be minimized if warehousing develops on a large scale.
- Warehouses render various subsidiary services, such as sorting and packing commodities for shipment, cleaning and drying goods and preparing them for the market, acting as forwarding agents for exporters of good, purchasing goods on behalf of clients, and collecting and disseminating marketing intelligence.

Progress

Storage and warehousing facilities for farmharvests on a commercial basis are available both in the public and the private sectors. The public sector dominates and accounts for a significantly larger share of the total capacity available in the economy.

The main institutional agencies providing these facilities are summarised here as Central and State Warehousing Corporations, the Food Corporation of India and the cooperatives.

The Central Warehousing Corporation was set up in 1962 to do the following functions:

- Acquire and build godown and warehouses at suitable places,
- Arrange facilities for the transport of farm produce,and inputs,
- Subscribe to the share capital of State Warehousing Corporations, and
- Act as agent of the Government for the purpose of purchase, sale, storage and distribution of farm produce and harvests.

The State Warehousing Corporations have been set-up at places of the state and district importance. Their functions are the same within a state as those of CWC at the national level.

Ideal Marketing Method

The ideal marketing method is one that maximizes the long run welfare of society. To do this, it must be physically efficient, otherwise the same output could be produced with fewer resources, and it must be effectively efficient, otherwise a change in allocation could increase the total welfare and where earnings distribution is not a consideration.

For maximum physical efficiency, such basic physical functions as transportation, storage, and processing should be carried on in such a way so as to achieve the highest output per unit of cost incurred on them. Similarly an ideal marketing method must allocate farm products in time, space and form to intermediaries and consumers in such proportions and at such prices as to ensure that no other allocation would make consumers better off. To achieve this condition, prices throughout the marketing method must be efficient and must at the same time be equal to the marginal costs of production and marginal consumer utility.

The following characteristics should exist in a good marketing method:

1. There should not be any government interference in free and market transactions. The method of intervention includes, restrictions on food grain movements, restrictions on the quantity to be processed, or on the construction of a processing plant, price supports, rationing, price ceiling, entry of persons in the trade, etc. When these conditions are violated, the inefficiency in the market method creeps in and commodities pass into the black market. They are not then easily available at the fairest prices.
2. The marketing method should operate on the basis of the independent, decisions of the millions of the individual consumer and producers whose lives are affected by it.
3. The marketing method should be capable of developing into an intricate and far-flung marketing method in view of the rapid development of the urban industrial economy.
4. The marketing method should bring demand and supply together and should establish an equilibrium between the two.
5. The marketing method should be able to generate employment by ensuring the development of processing industries and convincing the people to consume more processed foods, consistent with their tastes, habits and earnings levels.

Scientific Marketing of Farm Products

The tendency among the farmers to market their produce has been increasing. Production is complete only when the produce is marketed at a price remunerative to the farmer.

Increasing specialisation in production of higher marketable/ marketed surplus of the produce and alternative channels of marketing have increased the importance of the marketing activity for the farmers. However, marketing activity should be guided by certain basic principles which are briefly explained. The farmers can gain more if they follow the following principles of scientific marketing.

Always Bbring the Produce for Sale After Cleaning it

Impurities, when present, lower the price offered by the traders-buyers in the market.

The fall in price is more than the extent of impurity present in the produce would warrant. Clean produce attracts more buyers.

Sell Different Qualities of Products Separately

The produce of different varieties should be marketed separately. It has been observed that when different varieties of products are marketed separately, the farmers get a higher price because of the buyers preference for specific varieties.

Sell the Produce After Grading it

Graded produce is sold off quickly. The additional earnings generated by the adoption of grading and standardisation is more than the cost incurred in the process of grading and standardisation. This shows that there is an incentive for the farmers for the production of good quality products.

Keep Abreast of Market Information

Price information helps him to take decisions about when and where to sell the produce, so that a better price may be obtained.

Carry Bags/Packs of Standard Weights

Farmers should weigh their produce and fill each bag with a fixed quantity.

The majority of the farmers does not weigh their produce before taking it for sale and suffer loss by way of a possible malpractice in weighing, or they may have to make excess payments in transit (octroi, transport costs, etc.).

Avoid Immediate Post-Harvest Sales

The prices of the produce touch the lowest level in the peak marketing season. Farmers can get better prices by availing of warehouse facilities existing in their areas. Farmers can meet their cash needs by pledging the warehouse receipt to nationalised banks.

Patronise Cooperative Marketing Societies

Farmers can get better prices by sales through a cooperative and marketing society and can avoid the possibility of being cheated. The cost of marketing particularly the transportation cost for farmers having a small quantity of marketable surplus, is minimized, for transportation is arranged co-operatively by the society and the profit earned by the society is shared among its members.

Sell the Produce in Regulated Markets

The farmers should take their produce for sale to the nearly regulated markets rather than sell them in the village or unregulated markets. In regulated markets marketing charges are on very few items. They get the sales slips in the regulated markets, which show the quantity of the produce marketed and the amount of charges deducted from the values of the produce. Sales slip protect farmers against the malpractices of deliberate erroneous accounting or unauthorised deductions.

Conclusion

A good marketing method is one, where the farmer is assured of a fair price for his produce and this can happen only when the following conditions are obtained.

- The number of intermediaries between the farmer and the consumer should be small;
- The farmer has proper storing facilities so that he is not compelled to indulge in distress sales,
- Efficient transport facilities are available,
- The malpractices of middlemen are regulated,
- Farmers are freed from the clutches of village financiers and
- Regular market information is provided to the farmer.

3

Agriculture Production and Marketing

Improvements in the Production and Marketing

Marketing projects were limited to products which were subject to common price and market regimes but, as the range of commodities within the price policy expanded, this was no obstacle. Article 12 provided an interpretation of the four categories listed in Article 11 and it is interesting, even at such an early stage in the history of the CAP, that 'the adaptation and guidance of production' meant 'the quantitative adaptation of production to outlets' and 'improvements in the quality of the products'. During the 1960s—and indeed for a large part of the 1970s—Reg. 17/64 provided the main source of funds for structural reform. The range of activities covered by the Regulation was very wide, encompassing production and marketing projects, and those which were a mixture of the two. Over the lifetime of the Regulation, projects to improve production structures received 53 per cent of the funding, marketing structures 41 per cent, and mixed projects 6 per cent.

The Community contribution to the investment was initially very low at 25 per cent for all projects but, in 1973, this was raised to 45 per cent of eligible expenditure in projects related to production, with the beneficiary only having to provide a minimum of 20 per cent of the cost (originally a minimum of 30 per cent). Thus, right from the beginning, the Community element in structural legislation was low, unlike the price and market policy where, by the end of the 1960s, FEOGA-Guarantee was providing 100 per cent of the support. However, it should be noted that, for structural measures generally, the tendency over time has been for the Community contribution to rise, particularly in the poorer regions or in the poorer Member States.

During its lifetime, Reg. 17/64 was extremely popular and far more projects were submitted for funding each year than were ultimately aided. The sectoral distribution of aid, 1964-1979. Three sectors —milk, wine, and meat—accounted for one-third of the aid; land improvement and hydraulic works accounted for about another third; and all other purposes the remainder.

The result was that the better-off farmers and the more prosperous farming regions were benefiting disproportionately from all aspects of the

CAP. An interdepartmental group within the Commission reported *inter alia* on this issue, and its findings in relation to the regional impact of Dir. 72/159 have already been discussed.

As to the regional impact of Reg. 17/64, if it had been operating as originally intended, each Community Programme would have had specified zones of principal action within which investment aid would have been concentrated. The Commission had estimated that, as a general rule, these zones would have represented about one-third of the designated area or other appropriate measure, such as volume of production.

In reality, in Reg. 17/64, there was no provision for regional differentiation of aid, rather the funds available were allocated to the Member States according to a prearranged system of quotas. Thus, prior to Enlargement, Italy received annually an allocation of just under 34 per cent of available credits, Germany 28 per cent, France 22 per cent, and so forth. These percentages were adjusted downwards to accommodate the three new Member States after 1972.

The interdepartmental group was interested to examine how these national quotas had operated in practice and to compare them with the percentage distribution among the Member States of: (a) the agricultural labour force; (b) final agricultural production; (c) the simple average of (a) and (b), which in effect gave production weighted by the inverse of agricultural labour productivity; and (d) agricultural area. In no case did the quota exactly match any of these measures, although it came reasonably close to (c). For some countries the divergence was considerable—which was hardly surprising, as the quotas had been devised more on a basis of horse-trading than on any economic or social indicator.

Within some of the Member States, the group discovered that the quotas were apparently being divided by region, according to the size of the agricultural area. This meant that 'some of the relatively extensive and well-structured agricultural regions of the community (such as Schleswig-Holstein, Niedersachsen, Scotland, Emilia-Romagna, Toscana and Lazio) receive much higher expenditure than agricultural employment or structural criteria would suggest'.

The group also commented on the considerable difference between expenditure commitments for Italy and actual payments: in 1964-72 it received about one-third of total commitments but only half that amount had been paid by 1974. Thus, even at this early stage, administrative difficulties were preventing Italy from gaining access to structural funds—a pattern that was to become familiar as the years went by.

It is of some interest to examine whether or not the perceived lack of balance between the need for structural improvement in poorer regions and the distribution of total payments from Reg. 17/64 persisted over the lifetime of the scheme. In 1989, the Commission published a cumulative regional breakdown of the aid granted. While Italy received the largest payments, and while the South and the islands received a considerable share of the total, Emilia-Romagna remained a main beneficiary, and Lazio received more than Sicily.

Germany, which received the second highest payments, still had Niedersachsen as its second highest recipient, although Schleswig-Holstein did not maintain its earlier position. In the UK, Scotland remained by far the largest recipient. In Belgium, the south is poorer than the north but received less aid, while in Ireland two regions—South-East and South-West—dominated, although the West and North-West were much poorer.

Member State of the total number of projects financed. Most were already completed by the end of 1992. About 11.5 per cent had been abandoned, for whatever reason, by far the greatest number of these were in Italy. Despite its late arrival, the UK made very good use of Reg. 17/64. Member State of the aid granted, cumulatively to the end of 1992. This should be compared with the payments made. While the sources are not the same, which may have resulted in some discrepancies, in broad terms one can conclude that the account is nearly closed on this scheme. In Italy there is a considerable difference between aid granted and payments received, which presumably is a reflection of the large number of projects which were abandoned.

So the life of this extraordinary piece of legislation draws to a close. For so many years, it provided a link with the earliest days of the CAP and, in the distortions which so quickly developed in its operation, was a reminder of how soon in the life of the policy the politics of the CAP usurped the ideals of the original designers. However, some aspects of Reg. 17/64 have remained. In 1976, the interdepartmental group referred to earlier recommended that, when Reg. 17/64 ceased, its infrastructural aspects should be restricted geographically in any replacement legislation.

In a sense, this did occur with the measures which came into force in the late 1970s and early 1980s. However, it was not until the reorganization of the Structural Funds in the late 1980s that a more systematic approach to regional concentration of effort was introduced into the CAP. The aid to processing and marketing was carried forward with the enactment of Reg. 355/77.

The Establishment of Producer Groups

One of the most obvious features of agricultural production in Western Europe (and in many other parts of the world) is its organizational structure, dominated by small and medium-size independent farms. Production in such a framework comes very close to the economist's definition of a perfect market. Farmers are price-takers: they are very rarely in a position to dominate the market, even if they are organized into a commodity or special interest group. In general, they produce a homogeneous product and therefore cannot benefit from higher prices achieved through product differentiation. They are in a weak bargaining position, whether selling a product for direct consumption or for processing, and this is particularly true when the product is perishable (as with milk) or perishable and seasonal (as with fruit and vegetables).

The poor structural organization of agriculture was even more pronounced in the 1960s in the new Community than it is in the 1990s. Farming was characterized by a very large number of exceedingly small producers, each with an extensive range of products. This situation was not conducive to orderly marketing, particularly at a time when consumer demand was changing rapidly due to the marked increase in living standards.

What was required was regular supply, known and constant quality, and stable prices. It was essential that producers respond and that they took greater responsibility for the marketing of their own products. Vague references were made by the Commission as far back as the 1959 and 1960 proposals on the CAP to its intention to encourage initiatives by farmers' organizations, which would lead to better information on markets, to improve their stability, and their response to consumer demand.

The Commission reasoned that, if producers of the same commodity came together in an alliance, it would enable them collectively to improve the quality of their product by enforcing grading standards, and the continuity of supply, by having centralized storage and processing facilities. The group would market the output of its members, thereby improving their bargaining power and, as a result, obtaining a higher unit price. This in turn would help to further the aims of Art. 39.1 of the Rome Treaty, in particular as regards incomes, market stability, and supply availability.

These thoughts were carried a stage further in Reg. 26/62 on the application of certain rules of competition to production of and trade in agricultural products. Under this Regulation, farmers and their organizations were, in general, excluded from the ordinary Community rules on competition, in so far as their activities were concerned with the production and sale of agricultural products, and the use of common facilities for the storage and processing of such products.

Although, theoretically, the case for collective action is strong, in practice a recurring difficulty with any sector dominated by small-scale independent producers is to persuade them to join a group which will have power to impose standards, control prices, and even production, and once they have joined to prevent them from breaking their contractual obligations to the group, thereby undermining its collective bargaining power.

One way in which the advantages of membership of a producer group can be enhanced is to give to such groups certain formal market intervention responsibilities. Prior to the establishment of the European Community, the Netherlands provided good examples of how such a system functioned, especially in the fruit and vegetable sector. The characteristics of the Dutch system were outlined in the 1960 proposals on the CAP, and certain features of that system found an echo in some of the commodity regimes. The first example came in Reg. 159/66 *concerning additional arrangements for the common organization of the market in the fruit and vegetable sector*. This Regulation completed the arrangements for fruit and vegetables which had been hanging fire since 1962. A novel feature of this regime is the important role played by producer groups which act as the agency for internal market support.

For many commodities, provision is made in the relevant market regime for the withdrawal of produce from an oversupplied market, its storage, and disposal. These tasks are normally undertaken by an official agency or agencies established in each Member State. However, for a small number of products, market management is in the hands of the producers themselves, through producer groups which receive official recognition. Given the characteristics of fruit and vegetable production—a highly perishable product, many small-scale producers, local markets, seasonal production, volatility in supply and price-flexibility and speed of response are required if any form of market management is to be effective.

Apart from market support for certain commodities, more generally producer groups provide their members with facilities for handling and marketing the output of the relevant commodity. In their turn, the members are expected to market this total supply through the group (unless the group allows otherwise), and they must abide by the rules adopted by the group for the improvement of quality and the adaptation of supply to demand. The involvement of FEOGA-Guidance lies in the financial encouragement it provides for the establishment of producer groups.

The first example of such aid appeared in Reg. 159/66 under which Member States were given the possibility of helping producer groups to get started, by providing them with a degressive aid during the three years following their establishment. This annual aid was not allowed to exceed 3, 2, and 1 per cent respectively of the value of the production marketed by the group. The Guidance Section reimbursed 50 per cent of the Member

States' eligible expenditure under this scheme. Tentative references were made in Reg. 122/67 *on the common organization of the market in eggs* and in Reg. 123/67 *on the common organization of the market in poultrymeat* to the need to encourage efforts to improve the organization of production, processing, marketing, and quality through trade groups. However, in these commodity regimes, no financial support was offered to help launch such groups.

In the light of all this piecemeal and unco-ordinated interest in producer organizations in various commodities, it is not surprising that the Commission eventually proposed the establishment of a Community framework within which Member States could encourage the formation of producer groups in any commodity. The Commission was also prompted to do so because certain Member States were already promoting these groups and there was a danger that problems would arise at Community level through the use of different criteria and objectives in the various countries concerned.

Thus, in 1967, the Commission proposed a Regulation on producer groups and their unions in COM(67)68 final, which set out common rules on the type of aid which could be given by the Member States to encourage the formation and development of these groups, and on the conditions under which they could be officially recognized. Despite the Commission's evident interest in the creation of producer groups and despite the existing Community launching aids in the fruit and vegetable sector, the 1967 proposal did not make provision for Community aid. This was somewhat surprising and indeed was criticized by the European Parliament in its Resolution on the draft.

Although Mansholt always regarded producer groups as an extension of market organization rather than as a part of structural policy, his 1968 Plan did contain a discussion of the need to improve marketing, and one of the measures suggested to aid that process was the immediate adoption by the Council of the proposed Regulation on producer groups. The Council clearly did not heed this exhortation because, in its 1970 structural reform proposals, the Commission made a pointed reference to the thorough discussions which had taken place in the European Parliament and in the ESC on the 1967 producer group draft and stated that the Council had not even begun a discussion on it.

The 1970 reform proposals included a considerably revised draft Regulation on producer groups—the most important change being the provision of Community aid to help launch the groups. In the explanatory note to the proposals, the Commission made it clear that it regarded the improvement of market structure as primarily the responsibility of the Member States. For that reason, it proposed that the Community should refund only 30 per cent of eligible national expenditure. The logic of this position is unclear, as it will be recalled that the Community was already refunding 50 per cent of eligible expenditure on producer groups in the fruit and vegetable sector.

In 1971, when the revised structural reform package was presented to the Council, the draft Regulation on producer groups appeared again—now in a third version. The changes made were generally of a technical nature but, importantly, the Community contribution to eligible expenditure by the Member States had been reduced to 25 per cent. It was all to no avail, as the draft Regulation was not passed by the Council.

In the *Resolution* on the new orientation of the Common Agricultural Policy in 1971, the Council finally agreed in principle to the introduction of a regime to encourage the formation of producer groups but, once again, it appeared that Community aid would not be offered and that the funding would be by the Member States. Despite this, however, the Council remained unable to agree to the proposal, although discussions dragged on into 1972. The only progress which was made was under Reg. 1696/71 *on the common organization of the market in hops*.

The hops producer groups are not involved in price support operations but are intended in a more general way to assist in the centralization of supply and in the adaptation of the product to market requirements. Member States were free to grant a launching aid and FEOGA-Guidance reimbursement was limited to 25 per cent of eligible expenditure. An extra feature of this regime was the option given to Member States to aid producer groups on a temporary basis to reorganize hop gardens and to switch to more suitable varieties. The form of aid was a payment per hectare replanted or reorganized, and FEOGA-Guidance refunded 50 per cent of the eligible expenditure of participating Member States.

Discussions resumed in 1976 on the introduction of a general producer group regime and, in the following year, the Commission put forward yet another revised proposal. This was so radically different from previous drafts as to be, in effect, a completely new proposal. In the explanatory memorandum, the Commission tried to explain why such a harmless piece of legislation was causing such difficulties.

Basically the problem was that certain Member States with well-developed production and marketing structures could not see the necessity for this legislation. If anything, the differences between the Member States had grown since the original 1967 proposal. The Commission, therefore, proposed that instead of a Regulation of universal application, there should be a measure which would apply only in those areas where the supply of agricultural produce was structurally very inadequate. Primarily this meant Italy and indeed the proposal referred solely to that country, with the provision that the Regulation could be extended to other Member States.

The Council had agreed in February 1977 that it would adopt the proposal by the end of June. It was in fact a full year later that Reg. 1360/78 *on producer groups and associations thereof* was enacted. It was almost totally unrecognizable from the draft and covered not just Italy but also Belgium, Southern France, and DOM (the French overseas departments). The commodity coverage differed by country, and in France by region. It was intended to be of assistance to farmers in regions and sectors where there were severe structural deficiencies, due to the small size of farms and poor organization. For a measure which was to last initially for five years only, at a total FEOGA cost of 24m. ua, an incredible amount of bureaucratic effort was involved. When Greece, Portugal, and Spain joined the Community, Reg. 1360/78 was extended to them, and to Ireland and the remainder of France in 1988, all for specified lists of commodities.

The only other commodity for which producer group launching aids are available is cotton under Reg. 389/82. Under the terms of Accession, when Greece joined the European Community, market support was extended to cotton and producer groups were seen as an important mechanism for the rationalization of production and the centralization of handling and marketing. For this reason, the Regulation covers not only launching aids but also investment aid to assist with structural improvements in harvesting, ginning, storage, and packaging. Producer groups in Italy and Spain are also eligible for aid under this regime.

Having sketched the rather confused history of the development of legislation on producer groups, the remainder of this Section reports on the statistical evidence of the success or failure of the various structural measures to persuade farmers to come together to improve the organization of particular commodities. Data on structural aids are available for producer groups in fruit and vegetables, hops, cotton, and miscellaneous products under Reg. 1360/78.

In 1972, Reg. 159/66 was replaced by Reg. 1035/72, which consolidated all the existing Community legislation on fruit and vegetables. Arrangements for launching aids for producer groups were retained unchanged. The pattern which is so familiar from other structural aids is present here also: aid is concentrated in a very small group of countries and it is not necessarily related to the size of the sector. For instance, the very small number of producer groups set up under this legislation in Belgium and the Netherlands is indicative of the strength of the previously long-established marketing structures based largely on the cooperative movement. Therefore, there was little incentive or need to seek launching aids. In contrast, the relatively large number of launching aids in the UK is an indication of the previous absence of joint ventures by farmers rather than an indication of the size of the horticultural sector. The size of producer groups varies considerably from one Member State to another, as evidenced by the absence of any relationship between the number of groups and the level of investment.

Hops provide a real contrast to fruit and vegetables. This is a very small sector, highly localized in traditional areas of production where, of course, it can be quite a significant crop. Much of the output is grown on contract. The schemes of aid came to an end some years ago and 1985 was the last year for which data were available. The situation with regard to cotton is quite different. Greece is the largest producer and so far (at the time of writing) was the only Member State to have made use of the launching and investment aids, although the Commission anticipated that Spain, which is the only other significant producer, might well begin to show an interest in this measure.

In 1991, the Commission reported to the Council on the operation of Reg. 1360/78. By the end of 1990, the Member States had granted recognition to 560 groups. The discrepancy is explained by 'the slowness of the Member States' administrations in drawing up implementing rules and requesting refunds'.

Despite the fact that Belgium was included in the scheme right from the start, no groups in the relevant production sectors (cereals, cattle, pigs, and Lucerne) had applied for recognition, even though the requirements had been made easier.

Indeed, the Commission stated that the Belgian authorities had reported 'that those concerned are uninterested in the very idea of producer groups and that, in the present circumstances; this is unlikely to change in the future'. For that reason, the Commission proposed that, when the life of the Regulation was prolonged, Belgium should be excluded. This was rejected!

Improvements in Processing and Marketing

The fact that the processing and marketing support activities of Reg. 17/64 were continued in Reg. 355/77 on common measures to improve the conditions under which agricultural products are processed and marketed. This Regulation was a long time in coming. The life of Reg. 17/64 was limited following the introduction of the financial Regulation 729/70, which laid down the basis for future assistance of a structural nature. If for no other reason than that of legal necessity, consideration had to be given to the replacement of Reg. 17/64 by some other measure in the sphere of marketing and processing.

More than that, however, there has always been a tendency to concentrate aid to agriculture in that sector of the industry which lies on the farm and to neglect the need to improve the handling of the products once they leave the farm. For many years, Reg. 17/64 was the only Community measure of any importance providing such post-farm aid, despite the fact that it was recognized that the agricultural sector was weak in the handling of produce.

In setting the new guidelines for the CAP in 1971, the Council had invited the Commission to continue to study the problems of marketing and processing, and to submit a proposal on their alleviation. In 1973, in the *Improvement* memorandum, the Commission announced its intention to make such a proposal but it was not until mid-1975 that it finally managed to submit this proposal to the Council in COM(75)431. The Commission reasoned that a more efficient processing and marketing sector would be in a position to pay better prices to producers, to diversify output, thereby stimulating demand, to concentrate more on exports, and (rather obscurely) to 'better handle products from remote areas of the Common Market'.

The proposal was for the establishment of multi-annual programmes, each of which would cover one or more agricultural products. Within these programmes the Community would provide aid for suitable projects intended to promote the rationalization and expansion of processing and marketing. Assistance was to take the form of a capital subvention of not more than 25 per cent of the value of the investment, with the beneficiary providing at least 50 per cent and the Member State participating in the funding as well. The estimated cost of the scheme was 400m. ua over a five-year period.

While largely in agreement with the intentions of the Commission, the ESC had a number of reservations.

For one, the Committee was of the opinion that the proposed Regulation was 'of direct concern to consumers, farm workers, and workers in the food industry and the distributive trade' but that insufficient regard was being paid to the interests of these parties.

The ESC also believed that the effectiveness of the measure could be improved by the formation of producer groups and urged the Council to adopt the long out standing draft Regulation on such groups. This point was echoed by the European Parliament.

Another point on which the two bodies agreed was their scepticism concerning the Community's ability to control the new measure in a manner which would avoid exacerbating the surplus problem. Both Parliament and ESC considered the proposed scheme as financially inadequate. The Parliament stated that 'the Commission's proposal represents a limited step which will result in a decrease in the total real amount of Community aid to be granted for the improvement of marketing and mixed production/marketing structures, and which will make no substantial contribution to reducing agricultural surpluses and limiting the need for intervention'.

The ESC commented that '80 million units of account per year is not enough to solve all the problems of improvement of agricultural product marketing and processing facilities, especially as costs will continue to be pushed up by inflation'. It was not until 1977 that a modified version of this proposal was passed by the Council and it is hard to believe that either the Parliament or the ESC would have regarded the changes as beneficial.

The programmes 'to develop or rationalize the treatment, processing or marketing of one or more agricultural products' had to be drawn up by each Member State and agreed by the Commission before applications for the funding of individual projects could be considered. These projects could be for a wide range of activities: the development of new outlets for agricultural products; lightening the burden on intervention by improvements to the structure of a market; assistance to regions having difficulty adjusting to some aspect of the CAP; improvement of marketing channels or rationalization of the processing of an agricultural product; improvement of quality or presentation... While the contribution from FEOGA-Guidance to eligible expenditure was fixed at 25 per cent, in later years this was modified in various countries and regions, *e.g.* in the Mediterranean, and the West of Ireland. Strange little schemes were added on from time to time as, for instance, a special measure to rationalize and improve slaughterhouses in Belgium.

The support arrangements for processing and marketing were overhauled in 1990, so as to bring them into line with the new arrangements for the Structural Funds.

However, in some Member States there have been shifts, with new regions coming to the fore, such as the North East in Ireland. There still was evidence that the poorest regions were not achieving the enhanced level of investment necessary for them to narrow the income gap. There were great differences by country, with one or two sectors in each dominating: meat in Belgium; fish in Denmark; cereals and wine in Germany; fruit and vegetables, and tobacco in Greece; fruit and vegetables, and meat in Spain; wine, and fruit and vegetables in France; meat in Ireland; fruit and vegetables, and wine in Italy; fruit and vegetables in Portugal; and fish and meat in the UK.

Taking the Community as a whole, fruit and vegetables, meat, and wine projects dominated in terms of number, and together they also accounted for over 57 per cent of the funds.

As with Reg. 17/64, the outstanding feature of the table is the relatively large number of projects abandoned in Italy—over 38 per cent of the total. The figure for the UK—although much lower—is also rather high. The second part of the table gives the cumulative figures for funds committed and actually paid. The rate of reimbursement is an indication of the efficiency of the programme in the Member States, both in terms of the take-up of the investment opportunities and of the national bureaucracies. Five Member States had achieved a payment/commitment level in excess of 70 per cent; France and the UK achieved levels in excess of 60 per cent; Ireland and Italy in excess of 50 per cent. Greece, Portugal, and Spain were well below these levels. While the last two countries had only recently joined the Community and therefore could be expected to be somewhat slow the same cannot be said for Greece.

In terms of total funds, aid to marketing and processing is one of the biggest sources of structural investment in agriculture—just as Reg. 17/64 had been before it. Together with funds spent on improving the efficiency of farms, and LFAs, these three are the most significant programmes. While aid to LFAs has a large social element in it, in that it is intended to compensate for inherent locational handicap, the other major aids are intended to make lasting improvements to the economic framework within which farming takes place and agricultural produce reaches the market.

Despite their popularity, one must question to what extent they have been successful in raising levels of efficiency, so that one could see a time when the need for palliative measures under the price and market policy would be reduced or even eliminated. If such a time is not yet in sight, then one needs to question whether the commitment to and content of structural policy within the CAP are adequate.

growth of commodity production in agriculture

The growth of commodity production in agriculture can be assessed by the extent of utilisation of inputs and the growth of outputs produced for the market. The 'green revolution', which was introduced by the World Bank in Third World countries in the mid-1960s, was part of the imperialist policy to penetrate the countryside for markets. Though irrigation may have increased, out of the 350 million acres of net cropped area, barely 50 million acres, or 14 per cent, is cultivated more than once a year.

If one crop takes on an average four months, 86 per cent of India's agricultural land remains idle for 66 per cent of the time per year. To put it differently, nearly 60 per cent of the cropped area is not at all used. If agriculture was to be put on a capitalist footing and capitalist farming was to predominate in India, there would not be such a gigantic wastage of land utilisation and such a slow growth of irrigation facilities. But besides the modernisaation, the number of wooden ploughs increased from 37.5 million in 1956 to 43 million in 1966 and further to 44.5 million (*i.e.* during the green revolution period) in 1972.

Credit

It is thus clear that the stranglehold of usurious capital still dominates all sections of the rural populace. But usurious is an indication of pre-capitalist, feudal relations of production, which sucks up the surplus and prevents it from seeking productive channels. Besides, this usurious capital is increasing day by day. Marx has said that, *"Usurer's capital as the characteristic form of interest bearing capital corresponds to the predominance of small scale production of the self-employed peasant and small master craftsman", "Usury centralises money wealth where the means of production are dispersed. It does not alter the mode of production, but attaches itself firmly to it like a parasite and makes it wretched. It sucks out its blood, enervates it and compels reproduction to proceed under even more pitiable conditions."*

Also with a large section of the peasantry having turned to HYV, (and now unable to turn back to the traditional varieties) and caught in a crisis with diminishing returns for their output, the peasants are dependent on large capital inputs each year which they are now unable to finance. This is forcing them to seek larger and larger loans. With the peasantry unable to even pay their interest on their cooperative loans, they are turning more and more back to the moneylender. This trend is bound to increase enormously in the coming years as the crisis in agriculture deepens.

Rate of Growth of Production

In fact, the annual rate of growth of production of all agricultural crops actually dropped in the second period; while it was 3.2 per cent in the 1951-52 to 1964-65 period, it was just 2.6 per cent in the 1964-65 to 1983-84 period. This was because the increase in area under crops increased phenomenally in the first period at an annual rate of 1.7 per cent, while in the second period the annual increase was just 0.4 per cent. In other words, much better results in agricultural production could have been achieved by increasing the area under cropping than by introducing HYV. The economic and scientific research foundation has estimated that soil erosion and deforestation has been so acute that India looses about 1 per cent of its cultivable land every year to deserts. It is said that out of a total of 306 million hectares of cultivable land, 145 million hectares are either threatened with erosion or badly in need of soil and water conservation measures.

Generally, it can be summed up that production and yield of cropping has increased, with small increase towards cash crops, indicating some growth in capitalist relations. This is indicated through the change in some pockets of agriculture to capitalist farming and growth in commodity relations in large tracts of semi-feudal production, which continue their semi-feudal existence with enhanced contradictions caused by this lopsided growth of commodity relations within it. Productivity trends continue to indicate a backward mode of production with large farms being the least productive. Also with the increase in value of output being nowhere commensurate with the increase in value of inputs, it is clear that the growth in commodity production is lopsided, with this growth merely facilitating the dumping of commodities of the imperialists and comprador big bourgeoisie into the rural sector without significant returns to the farmer. This has led to a new set of contradictions in agriculture which are bound to intensify.

Distribution of the Productive Services in Agriculture

The structural changes in the industries of any large economic system of any kind embrace an intricate bundle of extremely varied but inseparably linked phenomenon and processes. In summarised form they display many features of the main development trends of society's productive powers in any one historical interval or another. Their generalised characteristic is a high degree of abstraction, necessary for scientific analysis, uniting the most diverse types of man's productive activity in a single whole. As Marx wrote:

Production in general is an abstraction, but a sensible abstraction in so far as it actually emphasises and defines the common aspects and thus avoids repetition. Yet this *general* concept, or the common aspect which has been brought to light by comparison, is itself a multifarious compound comprising divergent categories. The indicators mentioned in the preceding chapters of the resultant development trends in the postwar capitalist economy also reflect a high degree of generalisation. Nevertheless they are extremely important for a combined study of the real course of events. A scientific methodology for analysing broad social processes, including economic ones, requires us without fail 'to proceed from concrete realities, not from abstract postulates'.

In this connection Marxism-Leninism has always attached paramount importance to developing methods of making qualitative, *i.e.* statistical, estimates of socio-economic processes. Lenin, stressing the need for a concrete approach to 104 study of the main forms of capitalism's development in industry, wrote that one must examine the development of any particular form, after bringing out its essence and distinguishing features, by means of 'properly compiled statistics'. He made the same demand in regard to agriculture, for study of which we needed 'a picture of the process *as a whole,* with all the trends taken into account and summed up in the form of a resultant'. Only in that way can the decisive tendencies of economic development be concretely determined and the main point in them separated from the chance and secondary.

In his article 'Statistics and Sociology', written in 1917, Lenin stressed that the most widely used, and most fallacious, method in the realm of social phenomena is to tear out *individual* minor facts and juggle with examples. Selecting chance examples presents no difficulty at all, but is of no value, or of purely negative value, for in each individual case everything hinges on the historically concrete situation. Facts, if we lake them in their *entirety,* in their *interconnection,* are not only stubborn things, but undoubtedly proof-bearing things. Minor facts, if taken out of their entirety, out of their interconnection, if they are arbitrarily selected and torn out of context, are merely things for juggling, or even worse.

Hence a problem of vital interest for the study of social systems of any size, namely that of analysing the interaction of the whole and the part, inevitably arises, because the universal cannot exist otherwise than 'in the individual and through the individual'.

This problem becomes particularly acute when we are trying to elucidate the dynamics of the long-term tendencies in the distribution of the productive forces within such a broad system as the world capitalist economy. In it the 'individual' is not in fact simply the national economy of each country (in which the universal is also manifested 105 in an infinite variety of partial cases and specifically national features of development), but primarily embraces various groupings of these countries, the most important aspects of whose economies are united by the most essential common features into major sub-systems of the world economy. In the present historical situation these sub-systems include, first of all, the developed capitalist countries and the developing countries, *i.e.* the former colonies and semi-colonies that have thrown off foreign domination and taken the road, within that economy, of anti-imperialist struggle for independent national development. It would be logical to begin our description of the changes in the postwar distribution of the non-socialist world's productive power with an analysis of the distribution of its main industries in the light of Lenin's law of the uneven economic development of capitalism.

products of agricultural biotechnology

The regulatory approach to the safety evaluation of plants developed using rDNA technology has evolved in the best interests of research scientists, industry and the general public. The agricultural products of rDNA technology, such as GM foods and crops, may require approvals from up to three regulatory agencies; the US FDA, the USDA and the US EPA, depending upon the characteristics exhibited by the GM plant, its proposed use and introduced traits. The same standards of safety are applied to all products regardless of the technology used in their development.

The US FDA is responsible for ensuring the human safety of all new foods and food components, including products developed using rDNA technology, under the Federal Food, Drug, and Cosmetic Act (FFDCA). The USDA evaluates the potential of a GM plant to become a plant pest following its environmental introduction under the Federal Plant Pest Act (FPPA). The US EPA evaluates pesticides, including plant systems modified to express pesticides (*e.g.* insect-protected or virus resistance), under the Federal Insecticide, Fungicide, and Rodenticide Act (FIFRA).

As a result, the expression of an insecticidal protein in a food crop would undergo review by the USDA, US EPA and US FDA; a GM food crop exhibiting a modified oil content would be evaluated by the USDA and US FDA; and a non-food horticultural plant developed using rDNA technology for any other purpose (*e.g.* flower colour) would be subject to review by the USDA alone.

In most instances, obtaining all necessary approvals for the commercialization of an agricultural crop developed using rDNA technology takes a decade or more. However, the exact amount of time required will depend on the need to confirm performance, to evaluate characteristics of the food, environmental effects, and to produce the required amount of seed before the product can be distributed and commercially grown by farmers. Up to five years of field trials (5-10 generations of plants) are required for the developer of a new plant variety to collect sufficient data to meet the reporting requirements of the USDA. An additional five months to two years may be required for the US FDA, USDA and/or US EPA to complete all necessary product consultations, reviews and approvals.

Approval for the first commercial planting of a GM food crop was not issued until 1995. Since 1995, more than 40 new agricultural crops developed using rDNA technology have received approval for commercial planting within the United States. In 1999, approximately 72 per cent of the total 39.9 million hectares (more than 98 million acres) of GM crops grown worldwide were planted in the United States. Herbicide-tolerant soybeans (54 per cent), Bt corn (19 per cent) and herbicide tolerant canola (9 per cent) accounted for approximately 82 per cent of the GM plants cultivated. With an increasing number of agricultural biotechnology products reaching later stages of commercial development, it is anticipated that the overall area planted with GM crops will continue to rise.

The regulatory approach to evaluating the human health and environmental safety of GM crops within the United States is best described as a science-based, case-by-case assessment of hazards and risks. This approach has provided the flexibility required to reduce the regulatory burden placed on products that have been determined to be of low risk or concern. All agencies involved in the regulation of plants developed using rDNA continue to implement and develop policies based on recommendations made within the Coordinated Framework. In the future, the US FDA, USDA and the US EPA will be dedicating additional resources towards communicating how GM food and food components are regulated within the United States, and how these regulations function to be protective of both human health and the environment.

food and food components

The US FDA is responsible for ensuring the safety and wholesomeness of all food and food components, including the products of rDNA technology, under the FFDCA. The US FDA has the authority for the immediate removal of any product from the market that poses potential risk to public health or that is being sold without all necessary regulatory approvals. As a result, a legal burden is placed on developers and food manufacturers to ensure the commodities utilized and foods available to consumers are safe and in compliance with all legal requirements of the FFDCA. In order to understand the regulatory approach followed by the US FDA in the safety evaluation of GM crops, it is useful to consider food and food safety from a historical context.

People had been consuming foods derived from agricultural crops for many years prior to the existence of any food laws or regulations within the United States. Based on this experience, agricultural crops have been accepted as being safe for consumption as food, without additional testing to demonstrate their safety.

As long as the new crop variety has exhibited similar agronomic properties, and an appropriate taste and appearance, it has been considered safe to consume. As a result, most foods consumed today, in particular whole foods (*i.e.* fruits, grains and vegetables) and conventional foods, have not been subject to any kind of premarket review or approval by the US FDA. Nevertheless, food scientists have a good understanding that many of the commonly consumed agricultural crops contain natural toxicants (*e.g.* tomatine in tomatoes, solanine in potatoes, cucurbiticin in cucumber, psoralens in celery, etc.). As a result, new plant varieties may be subject to routine chemical analyses to ensure that none of these substances is present at potentially harmful levels. This type of general approach has been used in assessing the safety of thousands of new plant varieties that have been developed over a number of decades of crop breeding without compromising the safety of whole foods.

Role

Consistent with recommendations within the Coordinated Framework, the US FDA considered existing provisions of the FFDCA to be sufficient for the regulation of foods and food components developed using rDNA technology. It was concluded that the scientific and regulatory issues posed by the products of rDNA technology were not significantly different from those posed by conventional products.

As a result, GM foods and food components have been subject to the same standards of safety as already exist for the regulation of other foods and food components under the FFDCA. In order to better communicate interpretations of existing provisions of the FFDCA as they relate to the safety evaluation of foods derived from new plant varieties, including the products of rDNA technology, the US FDA released a policy statement in 1992 entitled 'Statement of policy: foods derived from new plant varieties'.

The US FDA has considered the use of genetic modification (*i.e.* rDNA technology) in the development of new plant varieties to represent a continuum of conventional plant breeding practices (*e.g.* mutagenesis, hybridization, protoplast fusion, etc.), and as a result, the safety evaluation of all new plant varieties, not just those developed using rDNA technology, have been evaluated based on an objective analysis of the characteristics of a food or its components, and not on its method of production.

Definition and Scope of Bioengineered Foods

The recently proposed rule of the US FDA concerns 'bioengineered foods' which have been defined as 'foods derived from plant varieties that are developed using *in vitro* manipulations of DNA (generally referred to as rDNA technology)'. As a result, the proposed rule has a much narrower focus than the 1992 US FDA Statement of Policy.

The US FDA has explained the need for a change in emphasis based on their expectations that many of the new plant varieties exhibit a greater potential to 'contain substances that are significantly different from, or that are present in food at a significantly different level than before'. As a result, the substances present in foods and food components derived from new plants developed using rDNA technology are less likely to be considered GRAS, and as a result, will require pre-market approval from the US FDA.

Safety and Nutritional Evaluation

The primary objective of the safety and nutritional evaluation is to demonstrate that the food derived from a new plant variety is as safe or nutritious as foods already consumed as a part of the diet. For new plant varieties, including those developed using rDNA technology, a science-based approach is used to focus the evaluation on the demonstrated characteristics of the food or food component. The evaluation of a GM food or food component typically involves reviewing information or data on any newly introduced substances, the known levels of toxicants, as well as the nutritional composition of the plant following modification. Substances that raise safety concerns (*e.g.* toxicants, allergens) would be subject to more extensive evaluation, since both intended and unintended changes may affect the levels of toxicants and nutrients in a food following the modification.

Guidance for performing a safety and nutritional evaluation was provided in the 1992 US FDA Statement of Policy, in a series of flow charts and text that cover:

- The crop that has been modified;
- Source(s) of the introduced genetic material;
- New substances intentionally added to the food as a result of the genetic modification (*e.g.* proteins, but also fatty acids, and carbohydrates).

Documentation required to support the evaluation typically includes: the purpose or intended technical effect of the modification on the plant, together with a description of the various applications or uses, a molecular characterization of the modification including the identities, sources and functions of introduced genetic material; information on the expressed protein products encoded by introduced genes; information relating to the known or suspected allergenicity and toxicity of any expressed gene products; for foods known to cause allergy, information on whether the endogenous allergens have been altered by the genetic modification; information on the compositional and nutritional characteristics of the foods, including anti-nutrients; and in some instances, comparative results of feeding studies involving the foods derived from plants modified using rDNA technology and the non-modified counterpart.

In performing its evaluation, the US FDA is particularly interested in the identification of inherent toxicants, known or potential allergens, assessing the concentration and bioavailability of essential nutrients, the safety and nutritional value of any newly introduced proteins, and the identity, composition, and nutritional value of modified carbohydrates, fats and oils.

If additional questions of safety remain following this evaluation, further toxicological studies may need to be performed. It is recognized that absolute assurance of the safety of any food does not exist. As a result, the goal of the safety evaluation is to establish a reasonable certainty of no harm under anticipated conditions of consumption.

With this in mind, experience with the existing food supply has provided the basis for evaluating the safety of new food or food components. Both the Food Advisory Committee and Committee for Veterinary Medicine have been extensively involved in the development of approaches for the safety and nutritional evaluation of foods and food components derived from new plant varieties, including those developed using rDNA technology.

Multiplier effect on agricultural production

Drought has multiplier effect on agricultural production during the subsequent year also, due to (i) non-availability of quality seeds for sowing of crops, (ii) inadequate draught power for carrying out agricultural operations as a result of either distress sale of cattle or loss of life, (iii) reduced use of fertilizers as the investment capacity of the farmers decline, (iv) non-availability of raw materials in agro-based industries, and (v) deforestation to meet the energy needs in domestic sector as agricultural waste may not be available in required quantity. The Central Ministry of Agriculture (MOA) is responsible for implementation and formulation of national policies and programmes to achieve agricultural growth through optimum utilization of the land resources, water, soil, plant, fisheries, and livestock resources. Government of India implements the following agricultural related Schemes (whether Watershed based or Agro-climatic region based) in the country, which deal agricultural resources information for Planning and Development:-

- Agro-climatic Regional Planning (ACRP) Project
- Agro-Ecological Mapping Project of the National Bureau of Soil Survey and Land Use Planning (NBSS&LUP)
- All India Soil and Land Use Survey (AISLUS)
- Early Warning System of Agricultural Situation in India
- Forecasting of Agricultural output using Space, Agro-meteorology and Land based observations (FASAL) Project
- Land Records Computerisation Project
- National Agricultural Research Project (NARP)
- National Agricultural Technology Project (NATP) to strengthen research-extension-farmer (r-e-f) linkage

- National Watershed Development Programme for Rain-fed Areas (NWDPRA)
- Soil and Water Conservation Programmes
- Drought Prone Area Development programme
- Desert Development Programme
- National Wastelands Development programme
- Integrated Mission on Sustainable Development (IMSD) Programme.?

Trends in the distribution of production

The long-term trends in the distribution of production in the former colonial regions are primarily governed by the resultant tendencies of the development of a few of the biggest countries (as regards area, population, and natural 164 resources). The ratio of their economic capabilities is therefore important in a generalised description of intraregional processes, including the contradictory consequences of the 'demographic explosion' in the Third World. In Latin America, for example, GO per cent of the inhabitants and nearly two-thirds of the aggregate GDP of the whole region were concentrated in Brazil, Mexico, and Argentina.

Among the 'Big Three' production developed at the lowest rates in Argentina in the postwar years, but the comparatively slow dynamics of its population growth led to its per capita production, for example, exceeding Brazil's in the 60s and Mexico's in the early 70s. It is typical that the economies of several small countries (Uruguay, El Salvador, Paraguay, Haiti, Honduras, and others) wore below the average of the developing countries of South and SouthEast Asia as regards the same indicators in those years, not to mention of bulk of the emancipated countries of Africa. The examples given, whose number could be greatly extended, very clearly confirm that there has been a steady sharpening of the unevenness of economic development of the former colonies and semi-colonies in recent decades.

This irresistibly intensifying process, however, in no way refutes the need (in addition to the study of the features of the economic growth of each country) for a comprehensive systems approach to analysis of the summary trends of their interconnected development within the whole developing world.

Such an approach helps, on the one hand, to bring out the objectively operating trends in which the extremely contradictory and isolated processes and phenomena characterising the shifts in the distribution of the productive forces of the separate countries and their regional groups are ultimately reduced, and on the other hand, gives the possibility of better understanding and appreciating the specific features of their development. In this connection comparative analysis of the long-term changes in the industrial structure of the social production of countries in Asia, Africa, and Latin America is becoming of paramount importance. As these countries are industrialised this process will undoubtedly hec07iie more and more pronounced.

Its effect gives grounds for supposing that even before the end of this century the share of agriculture in the GDP of most of the developing countries of Asia and Africa may fall by 50 to 50 per cent and reach the level attained in Latin America al the beginning of the 80s. At the same time the proportion of agriculture in the GDP of the Latin American region as a whole will also probably diminish, gradually approaching the present average level of the industrial capitalist countries (around 4 per cent). An inevitable consequence of that will be the pushing of increasing numbers of the rural population into the cities.

Another very important trend determining botli the current structure of changes in the production of the developing countries and their probable future structure has become the general though not uniform rise-in the role of industry in their GPP. The scale of its growth, moreover, has been quite regularly higher in the economically more backward countries. Thus the proportion of industrial output in the aggregate product for the period considered in the graphs increased in the African region by 120 per cent (from 12 to 26 per cent), in the Asian region by almost 100 per cent (from 14 to 27 per cent), and in the Latin American region by almost a third (from 24 to 31 per cent).

As a result of the increasing imbalance of this growth, a trend towards a certain convergence of the significance and place of industry in the general structure of these regions' production has been noted in the postwar period. The proportion of industry in the GDP of countries in South and South-Easl Asia and in Africa was 50 per cent less than in Latin America at the beginning of the 50s; at the beginning of the 80s the gap had noticeably closed and was estimated now at 1:1.3 for the former and 1:1.2 for tho latter.

This trend will intensify, in all probability, in the future, and characterise one of the tendencies of developing countries' uneven economic growth. It can be expected that the weight of industry in the GDP of most Asian and African] countries will come close to tho present level in the leading Latin American countries in the next two or three decades, while the latter will reach the average level of the present-day industrial centres of capitalism.

These very general estimates provide the dnta needed, but only the initial data, for studying the significance and place of industry in the regional economies. They naturally smooth over tho vast differences in structure and sectoral dynamics of the industrial production of the separate regions and countries.

A more and more concrete analysis of the separate industries is a central task of the part that follows, but here we must bear in mind that these processes of industrial development in agrarian and raw material producing countries are largely linked with rapid growth of mining, whose output is mainly sold on the world capitalist market, and in fact is still really controlled to one degree or another by expatriate monopoly capital. The rapid growth of the mining of minerals, and in particular of fuel, has played no small role in the regions being considered in the postwar rise in the proportion of industry in the gross product. It would not be legitimate, however, to conclude from this that the steady raising of the role of industry in the gross product of developing countries is more or less wholly 168 linked with the working of minerals.

The rapid development of their manufacturing industries in recent decades has had an ever increasing effect, especially the development of heavy industry, the volume of whose production rose on the whole by almost 150 per cent in 1968–80.

At the beginning of the 80s the heavy industries of all developing countries were producing 30 per cent more output in value terms (in added value in 1975 prices) than the light industries. There is little doubt that their basic industries will also grow al faster rates in the next few decades, which will further a gradual raising of their role in industrial production.

For a comparative estimate of the long-term trends bringing out both the general and the particular in the production sphere of the main developing regions of capitalism, it is essential to compare the mean annual growth rates of their main industries (including per capita rates) over some considerable period. Because of the inadequacies of the economic statistics of developing countries it is very difficult to bring out these growth rates in summary form (in contrast to developed countries). The comparison can only be attempted on the basis of very approximate calculations, which allow us to characterise the most general trends of the sectoral shifts with a certain degree of accuracy. At the same time they can be used as a starling point of sorts for studying the specific features of the development of production in the separate countries and groupings in each of the regions.

In agriculture, which is still in a slate of deep, protracted depression in the developing world as a whole, Latin America and most of the countries of the Near and Middle East have displayed relatively faster growth, so that a line towards an increase in their share in developing countries' total production of farm produce has begun to show., with a corresponding drop in the role of South and South-East Asia and Africa.

As a result they have become further differentiated as regards per capita farm production. By the beginning of the 80s the Latin American countries on the whole considerably surpassed the countries of Asia and Africa in this respect. It is unlikely that there will be any closing of this gap in the near future. The fight to eliminate the extreme backwardness of agriculture will undoubtedly remain one of the most pressing socio-economic problems of most of the countries of the former colonial world in the coming decades. It is the slow development of their agriculture, compared with population growth, that is the decisive cause of the consistent tendency manifested since the war towards a widening of the gap in GDP per capita between the developing and developed countries of the capitalist world.

In the 60s and 70s, however, this trend began to be more and more actively countered by processes developing in most of the other sectors of the economies of the former colonial periphery of capitalism, and primarily in industry. The international statistics clearly indicate that the per capita growth rates of industrial production in economically most backward regions of capitalism's periphery have noticeably exceeded the industrial centres in per capita growth rales of industrial production. In the developing countries of Asia per capita industrial production increased by a factor of 4.3 from 1950 to 1978, in Africa by 3.5, and in the developed capitalist countries taken together by 2.6. The countries of Latin America also showed a rather faster growth in this respect, on an average, than the main economic regions of capitalism.

In present conditions there has begun lo be a tendency towards a certain closing of the gigantic, gap in per capita industrial production formed under colonialism, together with 171 a gradual increase in the weight of the agrarian and raw material producing countries in the industrial production of the non-socialist world. In Hie very early 50s this gap was estimated at 1:35 for Asian countries, 1:32 for Africa, and 1:5.5 for Latin America. By the end of the 70s it had been clearly reduced and was expressed by the following approximate ratios: Asia 1:24, Africa 1:27, and Latin America 1:5.3. In the middle of 70s the substantial slowing of industrial growth rates in the developed countries connected with the greatest postwar cyclic crisis of the capitalist world economy fostered a consolidation of this trend.

The general course of this process leads to th,e conclusion that it is acquiring a quite stable, irreversible character in Ihc present situation. Being a natural consequence of the break-up of the colonial system, it has already begun to have (and in all probability will continue to have) a growing impact on the steady deepening of the crisis of the unequal system of international capitalist division of labour built up on a colonial basis.

At the same time the facts adduced confirm the extreme complexity and difficulty of the problem of overcoming the age-old industrial backwardness of the former colonies and semi-colonies, especially with their constant exploitation by expalriate monopoly capital. The solution of such a major and objectively inevitable hislorical task will obviously require several decades and will continue for a long time to be one of the urgent problems ol the light of the peoples of the liberated countries for final liquidation of the heavy legacy of the colonial period and imperialist domination of their economies.

To reduce their backwardness by even two-thirds from the present mean per capita industrial production of the developed countries, for instance, it would take the countries of South and South-East Asia around a quarter of a century at the rates of real industrial growth of the 60s and 70s, Africa more than a quarter of a century, and Latin America around 30 years. Then, however, per capita output of industrial goods in each of these regions would the respectively one-eighth, a ninth, and about a third below the level already reached by Ihe advanced capitalist countries at the end of the 70s. If that level remains unchanged in the future, then in that hypothetical case the way out for most developing countries would have lo be a 15-fold or 16-fold expansion of the volume of industrial production within the present-day world capitalist economy. These figures cannot, of course, serve as a generalised criterion of a quantitative, let alone a qualitative, estimate of the future parameters and goals of the industrial development of all the developing countries of Asia, Africa, and Latin America. In fact, however, they help determine only the scale of the industrial backwardness of developing countries. It inevitably follows from them that the struggle for a cardinal solution of the problem in the present transitional epoch will give rise to an urgent need to develop a long-term strategy and-strictly scientific approach to the global outlook for industrialisation of the former colonies and semi–colonies on quite other principles of social and world economic relations compared with those historically built up by capitalism.

The postwar processes in the agrarian and industrial spheres have consequently had a different effect on the resultant ratio of per capita GDP in the groups of countries under consideration. In the agrarian sphere they have fostered a widening of the economic gap, and in the industrial sphere a narrowing. But since the role of the agrarian sector in the GDP of developing countries is more significant than in the centres of capitalism, and the industrial sector less significant, the gap between them in per capita output of the main sectors (agriculture, industry, and building) has also continued to widen in recent decades, and for Asian and African countries was more than 10:1 at the beginning of the 80s against 8:1 in the first postwar years. In other words, at the turn of the decade to the 80s, per capita production in value terms (in comparable prices) was much lower in developing countries compared with capitalist ones, than at the turn of the 40s and 50s.

Industry Growth in Developing Countries

As for industry, its growth in developing countries has still not led to any real qualitative shifts in the liquidation of age-old industrial backwardness. The struggle to lay an industrial foundation for their national economies is still more or less, as a rule, in the initial stage. In spite of a narrowing of the gap in per capita industrial output, they now produce dozens of times less product than the average for capitalist countries.

The most powerful, technically advanced branches of production, moreover, that can more efficiently exploit the fruits of the present scientific and technical revolution are concentrated in the latter. That is affecting the dynamics of productivity in the industry of both groups of countries. In the 60s and 70s its annual growth rates in the economic centres of capitalism were double that in the developing countries.

There has been a quite stable tendency towards a gradual, tliough slow, rise in the proportion of circulation and services in the GDP of the developing world, which rose perceptibly in 1900–80. As a result of the rapid expansion of this sphere its weight in the GDP of the Asian and African regions rose by a quarter and nearly a third respectively in the 50s through the 70s. It rose a little as well in Latin America on the whole and constituted more than half of that region's GDP at the end of the 70s. In other words, there was a certain convergence of structural proportions in the sectors of non-materialproduction of the main developing regions, but the historical underdevelopment of these sectors still remains a distinguishing sign of the economic backwardness of the vast majority of the former colonies and dependent countries of these regions.

The proportion of developing countries in the aggregate production of the services and circulation sphere of the world capitalist economy is beginning lo grow. At the start of llio 80s it was 10 per cent against 11 per cent in the early 50s. The per capita gap between them and the developed countries, it is true, not only did not in fact diminish but even widened a bit on the whole, and was estimated at 15.5:1 in the late 70s, *i.e.* was considerably wider than in the sectors of material production.

For the main sectors of the circulation and services sphere this gap has not built up in the same way; in practice it has always been narrower in trade. But it is in that sector, as in agriculture, that the capitalist countries continued steadily to outpace the developing countries in per capita indices of growth, so that, although the proportion of the latter in the total value of the home trade of the capitalist world lias risen since the war (from one-seventh to one-sixth), there has been another trend in per capita terms While the value of per capita trade (in constant 1975 prices) in the GDP of capitalist countries was 12 times as much as in developing countries at the beginning of the 60s, it was 16 times as much at the start of the 70s.

This trend above all determined the effect in the circulation and services sphere of the tendency towards a further widening of the general economic gap in per capita GDP between the two groups of countries. The intensification of this trend also to no small extent was promoted by processes taking place in the sectors producing so-called services.

Developing countries' backwardness in this respect has long been deeper than in the other sectors of the sphere of social production we are considering. At the turn of the 50s and 60s it corresponded approximately to the level of their industrial backwardness. At that time less than one-tenth of the value of the services created in all capitalist countries was produced in developing countries, and the gap per capita was nearly 20:1. In recent years there have been changes in various directions in these ratios. The rapidly increasing dynamics of the growth of services in developing countries has led to their share rising in absolute terms to one-seventh at the end of the 70s. But in per capita terms the gap continued to widen for a long time, and in the early 70s could be expressed as 19.5:1, which was mainly due to the backwardness in this 175 area of the economically less developed and more densely populated countries of the African and Asian regions.

In the second half of the 70s alone, especially during the world cyclic crisis of 1974–75, a long-term consequence of which was a marked slowing of the growth rales of services in the centres of capitalism, did this gap begin to narrow a bit; at the beginning of the 80s it was expressed by a ratio of 17.5:1 (in 1975 prices). In other words, it has remained wider than in industry. The extreme backwardness of the services sphere governs several of the paramount aims of the long-term socio-economic strategy being developed by developing countries today in their striving to attain the maximum possible speeding up of the rates of social progress in present-day conditions and to raise the standard of living of their populations.

The only major group of services and circulation sectors in which developing countries managed to get a relatively stable increase in the dynamics of per capita production over that of the developed countries in the postwar decades was transport and communications; the grand total of per capita production in them increased by a factor of 3.4 against 2.4 in the developed countries.

There was a corresponding ri se (from 11 to 16 per cent) in developing countries' weight in the aggregate value of the product of transport and communications in the world capitalist economy. In per capita terms their mean annual rates of development have been higher of late than those of the industrial regions, so that in the area of the infrastructure, and in industry, there has begun to be a relatively new and apparently promising trend towards a certain narrowing of the gap in per capita indices, though it still remains extremely wide. If the present dynamics of the growth of production and population is maintained the gap should be halved at least on an average in the coming two or three decades in each of the main regions of the developing world, but its ultimate closing will undoubtedly still remain an urgent task for a long time.

At the same time there was a clearly increasing unevenness in the development of transport and communications in the developing countries, which, as the figures given above show, 176 was organically characteristic of all the other sectors of tho circulation and.services sphere. Its dynamics will also undoubtedly be a close function of the degree and scale of development of material product ion and of the population grovyth rates in the various regions and countries of the developing world. On the whole, however, one can say with confidence that as the productive forces rise the weight of this sphere in the social production of these countries will gradually increase although it will still lag greatly behind the corresponding average indicator for the industrialised countries for several decades.

The wide gap between the developing countries and the centres of capitalism in all the decisive areas of economic activity consequently remains one of the objective realities of our time. Formed long ago, during the building of the world capitalist economy on a colonial basis, it now characterises many of the most important features of that economy, and of tho whole system of its postwar international relations.

The developing countries of Asia, Africa, and Latin America, given the existing differences in their levels of sociopolitical and economic development, arc inevitably coming up against a need to deal with several common problems, vitally important for all developing countries, on both an international and a national scale. They include the very intricate set of each country's national problems, but ultimately tho struggle to cope witli all of them, taken together is objectively directed to eliminating the pernicious surviv' als of the colonial period from the affairs of human society" and to emancipating the developing countries from neoco, lonialism and continuing capitalist exploitation.

At the same time, as will be clear from the facts adduced above, the gap in levels of economic development remains extremely wide, in spite of considerable progress by the developing countries in coping with these problems, and moreover retains a tendency to widen in several determinant indicators, above all in gross product per capita. The relations of inequality and rapacious exploitation of the natural and human resources of the developing countries by expatriate monopoly capital prevailing in the world capitalist economy are furthering maintenance of this trend, even since the break-up of imperialism's colonial system.

The capitalist social system itself not only bears responsibility for the socio-economic backwardness of the peoples of the former 177 colonial world, but the exploiter system of international division of labour created by it largely continues today to fetter their productive forces. The basic patterns of the world capitalist economy still operate in the direction detrimental to the developing countries.

The striving of the developing countries to spread the liquidating of colonialism to the economic sphere, to put an end to their exploitation by the industrial powers of the West, and to achieve creation of the necessary external and internal conditions for attaining a level of development in the foreseeable future corresponding to the needs of today, is quite justified. At the same time all the nations of the earth without exception have an interest in the speediest attainment of these aims. The struggle for a steady acceleration of socio-economic and cultural progress in the former colonies and semi-colonies is therefore becoming an inseparable part of the fight of all progressive mankind for detente and the consolidating of lasting peace, for further development of the scientific and technical revolution for peaceful purposes, and for a favourable solution of the other global problems of today on which the whole subsequent course of world development to an enormous extent depends.

In his day Karl Marx concluded that mankind thus inevitably sets itself only such tasks as it is able to solve, since closer examination will always show that the problem itself arises only when the material conditions for its solution are already present or at least in the course of formation.

In the present historical situation more favourable objective conditions are being created than ever before for the countries of Asia, Africa, and Latin America that have liberated themselves from the imperialist yoke, or are in the course of doing so, for a successful struggle to overcome the backwardness inherited by them from the past in all sectors of 178 the economy and oilier areas of social affairs. But every kind of obstacle is still being put in llio way of this bard, long, stubborn light with the world capitalist economy by the former colonialists and expatriate monopoly capital. In these conditions the developing countries are uniting their efforts more and more firmly in the anti-imperialist movement to establish a new international economic order.

The Soviet Union and other socialist countries are supporting these efforts in every way possible and demonstrating their solidarity with the progressive aims of the struggle of the peoples of the developing world for a better future. As the Central Committee's Report to the 26'th Congress of the CPSU stressed: the CPSU will consistently continue the policy of promoting cooperation between the USSR and the ncwly-freo countries, and consolidating the alliance of world socialism and the national liberation movement.

The long-term trends reviewed in this chapter thus not only help us better to understand some of the general results of the changes that have already taken place in the structure of the distribution of the productive forces of the two groups of countries of the postwar capitalist world, but also to distinguish sqveral objective premises for analysing the probable outlook for subsequent structural shifts leading to aggravation of the internal contradictions and crisis phenomena in the capitalist economy. Study of these trends obviously calls for a need to allow in every way for the effect on them of the spontaneously operating laws of the cyclic development of the capitalist mode of production.

Agricultural Production Forms into the Capitalist Political Economy

Another thrust in the neo-Marxian political economy literature, and one that is somewhat contrary, has been the argument that the differentiation of agriculturalists into the capitalist and working classes may be incomplete in the foreseeable future as farm and non-farm production become integrated into a single system incorporating different organizational forms of production. An early statement of this line of thought was that of the German Social Democrat and Marxist theoretician, Karl Kautsky.

Kautsky argued that what was crucial to understanding the evolution of agriculture in advanced industrial societies was not simply the dominant form of ownership of agricultural enterprises, but rather the functions that were served by the emerging organizational formsof agricultural production.

This approach was developed further by Mottura and Pugliesi in an historical analysis of small holdings farmed part-time in southern Italy and the functions of smallholder agriculture in contemporary decentralized economic development. The thrust of their argument was that while most agricultural production took place on farms organized along capitalistic lines, part-time farming served as a backup alternative for the workers in industrial plants located in rural areas. In times of industrial contraction and high unemployment, displaced workers with small farms could turn temporarily to subsistence production until industrial conditions improved, thus forming a reserve labour force.

This integration of agricultural and non-agricultural production spheres has been further elaborated by Bonanno, who examined the role of the state in fostering small farms as one strategy to mediate the interests of the conflicting social classes in the emerging social orders of the advanced societies, particularly Italy and the United States. In this light, various farm programmes to deal with problems in the agricultural sector can be seen in part as rooted in the legitimation function served by the continuation of small farms. Small farms are also important in the decentralizing industrial system fostered by state policy. In this system, industrial firms move to rural areas where labour is not unionized and where wages are low because many potential workers have small farms producing inadequate incomes and few alternative opportunities. Work is also increasingly "informalized" in cottage industry, piecework arrangements. In such contexts, small farms serve the function as "keeper of surplus labour" — providing, at the same time, a source of low-cost labour for industry and, in the face of the tenuous employment, a source of security for the members of households with small farms.

Wenger and Buck, building on this line of thought but especially on the earlier work of Andre Gunder Frank, take this argument another step by examining how exploitation and superexploitation (extracting more value from workers' labour than permits reproduction of that labour) from members of farm households are necessary and dynamic features of both advanced capitalist societies and developing societies. Production organized according to obligations of kinship ("domestic relations of production") link ("articulate") in various ways with production organized along capitalist lines so that value is transferred from the domestic sphere of production to the capitalist sphere of production.

This takes place in such a way as to make the domestic sphere an "interstitial domestic reserve of labour" that subsidizes the capitalist sphere either directly or indirectly through various mechanisms. For example, unpaid household labour reduces both the wages necessary for workers employed in industry and the prices of agricultural commodities required by farm families. Off-farm income from wage work helps to pay the costs of agricultural production and thus lowers the price of food for other working class families.

Farmers as Actors in a Capitalist Political Economy

Whereas the perspectives of Mann-Dickinson and Friedmann on one hand, and of de Janvry, Friedland *et al.*, and others working in the Lenin (and, to a lesser extent, the Kautsky) traditions on the other, are, in a sense, in diametrical opposition, much of the most provocative work in the Marxist tradition in the "new sociology of agriculture" has revolved around formulations that explicitly or implicitly are attempts at synthesis. One of the most noteworthy attempts has been by Mooney.

Mooney's principal contribution has been to cast doubt as to whether the existence of conventional capital-labour relations on farms is an adequate benchmark for gauging the existence of capitalist penetration of agriculture. Following Wright, Mooney has developed a model of agrarian class structure involving "contradictory class locations" such that class positions other than family labour farmer (unity of capital and labour in the farm household), capitalist farmer, and agricultural wage labour are seen to exist. In particular, Mooney sees that there are several "detours" that can be taken by farmers in order to avoid proletarianization (*e.g.*, either being forced from agriculture or becoming hired agricultural laborers). These detours involve tenancy, contract farming, part-time farming, and debt. In each, whereas there is no capital-labour relation at the point of agricultural production, farmers are exploited by some fraction of non-agricultural capital (in tenancy, by landlords; in contract farming, by agribusiness; in part-time farming, by off farm capitalists; and in debt, by finance capital). Thus, Mooney argues that the exploitation of farm wage workers by agrarian capitalists is only one form that capitalist penetration of agriculture can take.

Moreover, Mooney sees that these detours — that is, the alternative ways in which capital acts to "strip simple commodity producers of... surplus value other than through the extension of wage labour" — may be more significant than full-blown capital-labour relations at the point of agricultural production. It is significant, however, that Mooney's explanation of why these contradictory class locations in agriculture tend to emerge has a subjectivist component, based on Weber's distinction between formal and substantive rationality. Mooney sees that many farmers are motivated more by forms of substantive rationality (*e.g.*, the desire for autonomy over their work) than they are by formal capitalist rationality. Accordingly, these farmers tend to be tenacious in holding onto their farms and farm lifestyles and will often tend to take one of the four "detours" to capitalist development in order to remain in agriculture.

Mann and Dickinson, however, replied vigorously to Mooney's neo-Weberianism with two major arguments: first, that so-called contradictory class locations are really based on the same social relations of production grounds that Mann and Dickson use and, second, that Mooney has misconstrued their notion of "obstacles to capitalist development" in agriculture to be immutable barriers, so that Mooney was unable to recognize that his notion of "detours" to capitalist penetration of agriculture is similar to their own views. Mann and Dickinson thus argue that subjectivism, and the explanatory ambiguities it involves, does not yield insights beyond those afforded by a structural, neo-Marxist theory. They were also critical of Mooney's project of synthesizing Marxian and Weberian approaches on the ground that these approaches are incompatible, except on an ad hoc, eclectic basis. Mooney added further to the debate by arguing that Mann and Dickinson have exaggerated his departure from a neo-Marxist framework-that his perspective retains Marxist insights by incorporating key Marxian categories as ideal-types within a Weberian framework.

He argued that this approach overcomes the incompatibilities of the Marxian and Weberian frameworks — unless, of course, one adopts a mechanistic Marxist or Leninist perspective. His key position was that the presumption of a predictably patterned and necessary differentiation of farmers into the capitalist class and proletariat is erroneous and that subjectively meaningful human action affects the process of differentiation substantially.

With the 1980s farm crisis coming under increasing scrutiny by rural sociologists, Mooney has extended his analytical framework to an historical examination of the state-sponsored farm credit system (Farmers Home Administration [FMHA], Commodity Credit Corporation [CCC], and the Farm Credit System [FCS]) and tax policy incentives for increasing the capitalization of agriculture. Mooney conceptualizes credit and tax policies as means of dealing with the crisis of legitimacy posed by the increasing prevalence of tenant farming produced by the Great Depression. Mooney argues that this credit and policy system created more indebted farmers who ultimately became vulnerable to financial crises in the form of rising real interest rates and declining agricultural commodity prices, a situation such as that which occurred in the 1980s. Financial crises expose to perceptive actors the role of the state in creating the crisis, thus creating another crisis of legitimacy and providing a basis for affected farmers to mobilize against the state.

Pfeffer has taken a considerably different tack than Mooney, though Pfeffer's work has some similarities to that of Mooney in that he stakes out middle ground on the question of whether the predominant feature of agriculture in advanced capitalism is the persistence of the family labour farm on one hand, or its demise into capitalist relations of production on the other.

Pfeffer conducted an historical analysis of three systems of agricultural production in the United States (industrial agriculture in California, family farming in the Great Plains, and the sharecropping system in the South prior to World War II) and demonstrated that recruitment and maintenance of access to a suitable labour force was a crucial factor shaping the disparate organizational structures of these three systems. Only in California, largely because of preexisting land concentration and of state actions to recruit and restrict the residential, educational, and occupational mobility of ethnic minority farm workers, was capitalist agriculture able to set root prior to World War II. Noting that the U.S. South had a high degree of land concentration during the entirety of the nineteenth century, Pfeffer goes on to explain why southern planters were unable to recruit a wage labour force after the Civil War and why the tying of sharecroppers to plots on the landlord's plantation through the crop-lien system was the only possible route to recruiting the labour necessary to operate a large plantation.

Finally, Pfeffer demonstrates that a combination of labour scarcity, the unwillingness of immigrants to work for wages on farms, and severe declines in the price of wheat in the late nineteenth century led to the establishment and persistence of family farming in the Great Plains. Pfeffer thus suggests that while capitalist agriculture can emerge under particular circumstances, there have been several alternative forms of agricultural organization — sharecropping and family farming, in particular — that have become established due to regional and commodity variations in the ability of farmers to recruit a dependable labour force.

Still a different approach was taken by Whatmore *et al.*, who have developed a theoretically based typology of farms in Britain. Their goal was to understand the effects of external pressures by non-family and non-farm capitals to penetrate agriculture and how these pressures articulated with internal processes on farms. By cross-classifying levels or degrees of subsumption of internal and external relations, they developed a typology with four main categories of farm types, ranging from relatively unsubsumed ("marginal closed unit") to the "subsumed unit." The internal relations variable pertained to ownership and control of farm capital and land, control over management, and the balance of family and hired labour.

The external relations variable pertained to level of dependence on industrial capitals for production inputs and purchase of farm outputs and to the degree of involvement in credit relations, especially with finance capitals. Applying this theoretically based typology to the data from three areas in southern England — an urban fringe area, an agricultural area undergoing commercial and industrial development, and a primarily agricultural area — they found that the process of subsumption of farm production relations, and thus their integration into the "wider circuits of capital," varied over time and space. Whatmore *et al.* concluded that farm families need to be understood as actors in the process of subsumption rather than its passive victims.

The neo-Marxian theoretical enterprise (and that of the modernization school as well) on the conversion of traditional peasant economic relations to commodity relations under capitalism has also been scrutinized by Vandergeest. Among his major criticisms were that neo-Marxist approaches tend to be too unilinear, too macro-structural, too denying of human agency, too inattentive to the role of the state in the process, and too unconnected with practice. He has argued, from a neo-Weberian position rooted in the work of Pierre Bourdieu, that all categories and theories are in the last analysis historically contingent, ideological, and interpretive social products.

The world is a complex whole which can only be investigated empirically and understood through theory. It cannot be reduced to the simple working out of a model derived through deductive theory — whether that of "simple commodity production" or of "capitalism". There is not a single deductive logic (such as the logic of exchange-value or accumulation) underlying or determining all relations in a capitalist formation, but there are different, historically contingent principles which we can only investigate through empirical research.

There has recently been a tendency in the neo-Marxist and neo-Weberian literatures towards a deemphasis on "deductivist"-functionalist theoretical postures. Subjectivist perspectives such as those of Vandergeest and Mooney have been influential in leading to this reorientation, though this tendency is by no means limited to those persuaded by neo-Weberianism.

The three postures in the Marxist political economy of agricultural tradition — those emphasizing the persistence of petty commodity production, those emphasizing the inevitable differentiation of petty commodity producers into antagonistic social classes, and those seeking synthetic positions — have yielded a continuing, lively debate in the rural sociology literature. Particularly instructive is the debate over the Mann-Dickinson hypothesis.

Also of importance is Goodman and Redclift's commentary on the work of Friedmann in which Goodman and Redclift critique Friedmann for taking an overly deductive approach to petty commodity production that ignores the role of historical conjuncture and ideology. Bernstein has made a comparable argument — that generic theories of simple commodity production under advanced capitalism tend to overemphasize its "functional" aspects and downplay contradictions and the diversity of household forms of production.

It should be emphasized that theoretical and empirical treatments of the political economy of North American agriculture have not been made only by rural sociologists. Many of the pioneers in this tradition of scholarship have been non-sociologists or sociologists who have had little or no connection with the Rural Sociological Society or the institutionalized form of rural sociology in U.S. land-grant universities. Non-etheless, rural sociologists have made influential contributions to this literature.

The emerging political economy tradition in the sociology of agriculture has led to a provocative and stimulating literature. It should be mentioned, however, that those working from this perspective are just beginning to establish a distinctive research programme; much of the existing literature, in fact, often tends to involve superimposing a new vocabulary on already-established data, such as Goss *et al.* set out to do. Establishment of a distinctive research tradition will probably occur slowly since conventional data sources such as censuses and sample surveys are often inapplicable to theoretical issues in the political economy of agriculture. There have, non-etheless, been several research papers in the political economy tradition that have utilized innovative new methodologies to generate research findings that bear directly on specific hypotheses.

Subculture and Agricultural Structure

It is useful to note that the growing interest in subjectivist approaches in the new sociology of agriculture has not been confined to those who work within the neo-Marxist and neo-Weberian traditions. Anthropologically-oriented researchers have also contributed to this literature through the examination of the effects of ethnic background on state and local level farm structure. Salamon and associates have done the pathbreaking studies in the anthropological tradition in Illinois communities, though recently this work has been taken up by others.

While their work was not intended to contribute to the "structure versus agency" debate that has raged between Mann and Dickinson on one hand, and Mooney on the other, and actually predated much of this debate, it is clearly relevant to the debate. In particular, the work of Salamon and associates pertains to the role of subjective motivations in farm decision making, especially motivations conditioned by persisting subcultural variations based in differences in ethnic origins of farm families.

The key finding of this line of research has been that farm families with German ethnic backgrounds tend to view farming as a way of life and hold a strong value for keeping the family farm intact. These Salamon has labeled "yeoman" farmers. In contrast, farm families with British ethnic backgrounds tend to be more entrepreneurially oriented, viewing farming as a way to make profits and having little attachment to farming or to particular farms. These Yankee farmers are more likely to seek growth in scale and less likely to support their local communities. This research has led to understanding differentials in farm size as a partial function of differences between these subcultural variants. Flora and Stitz found that yeoman farmers in a Kansas county generally expanded less than did Yankee farmers. Foster *et al.* also found support for the Salamon hypothesis in a study of a sample of Illinois farms that had been in the same families for 100 years or more.

Several other anthropological studies in the tradition of Salamon and colleagues have been reported in Chibnik. Of particular importance is Barlett's article in which she criticizes notions of the demise of the family farm and of the "disappearing middle" from an anthropological perspective.

4

Agricultural Finance and Management

Agriculture Finance

Finance in agriculture is as important as development of technologies. Technical inputs can be purchased and used by farmer only if he has money (funds). But his own money is always inadequate and he needs outside finance or credit. Professional money lenders were the only source of credit to agriculture till 1935.

They use to charge unduly high rates of interest and follow serious practices while giving loans and recovering them. As a result, farmers were heavily burdened with debts and many of them perpetuated debts. There were widespread discontents among farmers against these practices and there were instances of riots also. With the passing of Reserve Bank of India Act 1934, District Central Co-operative Banks Act and Land Development Banks Act, agricultural credit received impetons and there were improvements in agricultural credit.

A powerful alternative agency came into being. Large-scale credit became available with reasonable rates of interest at easy terms, both in terms of granting loans and recovery of them. Although the co-operative banks started financing agriculture with their establishments in 1930's real impetons was received only after Independence when suitable legislation were passed and policies were formulated. Thereafter, bank credit to agriculture made phenomenal progress by opening branches in rural areas and attracting deposits.

Till 14 major commercial banks were nationalized in 1969, co-operative banks were the main institutional agencies providing finance to agriculture. After nationalization, it was made mandatory for these banks to provide finance to agriculture as a priority sector. These banks undertook special programmes of branch expansion and created a network of banking services throughout the country and started financing agriculture on large scale.

Thus agriculture credit acquired multi-agency dimension. Development and adoption of new technologies and availability of finance go hand in hand. In bringing "Green Revolution", "White Revolution" and now "Yellow Revolution" finance has played a crucial role. Now the agriculture credit, through multi agency approach has come to stay. The procedures and amount of loans for various purposes have been standardized. Among the various purposes "Crop loans" (Short-term loan) has the major share. In addition, farmers get loans for purchase of electric motor with pump, tractor and other machinery, digging wells or boring wells, installation of pipe lines, drip irrigation, planting fruit orchards, purchase of dairy animals and feeds/fodder for them, poultry, sheep/goat keeping and for many other allied enterprises.

Schemes for Agriculture Finance

SBT Kisan Gold Card Scheme (General purpose Agriculture Term Loan)

Eligibility:

- Farmers having good track record of repayment for the last two years.
- Farmers who have closed their loan account without default and not our current borrowers.
- Farmers who have defaulted in repayment but closed the Loan within the stipulated repayment period.
- Farmers who are maintaining deposits with the Bank.
- Good borrowers of other banks provided they liquidate their dues with other banks.
- Good farmers who have not availed loans from any bank.
- *Purpose*: The borrower is at liberty to utilize 50 per cent of the amount for any purpose, including consumption purpose and purchase of land.
- *Amount of Loan*: The amount of loan is limited to five times the annual farm income including income from allied activities or 50 per cent of the value of the land offered as collateral security, whichever is less, subject to a maximum of ₹10 lakh.
- *Rate of Interest*: Interest rate ranges from 1 per cent below PLR.
- *Security*: Hypothecation of crops and assets, if any, created out of bank finance and existing movable assets such as milch animals, pump sets etc. The loan will be secured by equitable mortgage of properties worth double the loan amount, or term deposit receipts, LIC policies of adequate surrender value, NSCs completed lock in period or more etc.
- *Disbursement*: Cash disbursals are allowed to the full extent of the credit limit.
- *Repayment*: The repayment period shall be 10 years. The due date

of the installment shall be fixed in such away to coincide with the date of generation of income.

Homestead Farming

- *Purpose*: A scheme for financing farmers practicing mixed cropping/ intercropping along with allied activities to enable them to undertake cultivation of various crops in a more integrated way. The scheme provides the farmers with sufficient working capital required for their homestead farming (Mixed cropping along with allied activities) by fixing scale of finance based on land holding to meet the cost of entire farming activities.
- *Amount of Loan*: The farmers who own cultivated land below one acre be given the scale of finance on pro rata basis at the rate of ₹ 40000/- and farmers who own more than one acre of land be given at the rate of ₹ 37500/- per acre and part thereof.
- *Rate of Interest*: Interest rate ranges from 2.50 per cent below to 1.50 per cent above BPLR for various limits.
- *Repayment*: The facility will be sanctioned as an Agriculture Cash Credit limit (In case of Kisan Credit Card running cash credit).

Loan for Estate Purchase

- *Eligibility*: The estate should be either in yielding stage with the crops in its prime yield age or capable of being developed in to a viable unit. The yield/ net income of the estate should be sufficient to liquidate the proposed loan and interest accrued with in a period of 7 to 10 years. The proposed estate should be free from encumbrance and entire property should be offered as security to the loan.
- *Purpose*: To encourage those who prefer to settle down in agriculture and are in the look out of good/viable estates for purchase and also to improve production in agriculture.
- *Amount of Loan*: The quantum of loan that will be considered for sanction will be 75 per cent of the registered value or 50 per cent of the market value whichever is low. In exceptional cases 80 per cent of the registered value or 50 per cent of the market share whichever is low is also considered. The loan for the development of the estate like land development including working capital can also be sanctioned.
- *Rate of Interest*: Interest rate same as BPLR.
- *Repayment*: Repayment of loan will be in quarterly/half yearly/

yearly installments depending on the harvest of the crops and the loan shall be repaid within a maximum period of 7 to 10 years.

Kisan Credit Card Scheme

- *Eligibility*: All agriculturists who are in need of short term production requirements. ATM facility and Personal Accident Insurance Scheme for life up to ₹ 50000 and permanent disability cover up to ₹ 25000 is available on request.
- *Purpose*: To provide hassle free short-term credit to farmers on the basis of their land holdings for purchase of inputs and draw cash to meet their production needs. *i.e.* Cultivation expenses including allied activities with a consumption component.
- *Amount of Loan*: To be fixed on the basis of operational holdings and scale of finance with consumption component 15 per cent (maximum ₹ 10000/-) of production credit. The scale of finance to farmers who own cultivated land below one acre will be at the rate of ₹40000/- (on pro rata basis) and farmers who own more than one acre with intensive farming of land be given at the rate of ₹37500/- per acre and part thereof.
- *Rate of Interest*: Interest rate ranges from 2.50 per cent below to 1.50 per cent above BPLR for various limits.
- *Repayment*: Running Cash Credit account for 36 months subject to annual review and total annual credit should exceed annual debit.

Scheme for Financing Farmers for Purchase of Land for Agricultural Purposes

- *Eligibility:* Small and Marginal farmers - land maximum up to 5 acres of non-irrigated land or 2.5 acres of irrigated land including the land purchased under the scheme. Tenant, sharecropper and landless agricultural labourers with a good record of prompt repayment of our loans for the last 2 years are also eligible.
- *Purpose*: To finance small and marginal farmers, share croppers, tenant cultivators for purchasing land to expand activities and to make existing small and marginal units economically viable to bring fallow lands and waste lands under cultivation to step up agricultural production as well as productivity also to finance share croppers/ tenant farmers to enable them to diversify farming activities to allied areas to increase their income.
- *Amount of Loan*: Maximum loan under the scheme towards land

cost shall not exceed ₹ 5 lakh. Cost of development/economic activity shall be financed under the bank's other financing schemes.

- *Rate of Interest*: Interest rate ranges from 1.75 per cent below to 2.00 per cent above BPLR for various limits.
- *Repayment*: Repayment of the loan will be 7 to 12 years in half yearly/yearly installments with maximum of 24 months moratorium period. Gestation period/repayment due dates etc. will be fixed according to income generation from the activity.

Produce Marketing Loan (Advance against Warehouse Receipt)

- *Eligibility*:
  a. Farmers/ traders depositing farm produce in the warehouses of the central/ state warehousing corporations.
  b. Scheme will be operative in Karnataka, Andhra Pradesh, Tamilnadu and Kerala.
- *Purpose*:
  a. To protect the farmers from the compulsion to sell their produce immediately after harvest of produce despite an adverse market.
  b. To finance farmers and traders against warehouse receipt.
- *Amount of Loan*: 70 per cent of the value of the warehouse receipt, valued at the market value or 70 per cent of the market price advised by Agri. Dept, HO whichever is less.
- *Rate of Interest*

  Farmers:
  - Up to `3 lakh - 3.50 per cent below PLR 9.50 per cent
  - Above `3 lakh - 2.50 per cent below PLR 10.50 per cent

  Traders:
  - 2.50 per cent below PLR 10.50 per cent (Irrespective of the limit)
- *Repayment:* On demand/6 months which can be extended up to 12 months subject to satisfactory shelf life/market condition.

Agri-Loan to Non-Resident Indians

- *Eligibility*: Agricultural advances are available to the resident family members (means spouse, father, mother, brother, sister etc.) of Non-Resident Indians for land-based activities in respect of the land held by them in India subject to:
  a. The loan should be need based and the total land holding of

the Non-Resident Indian, in individual name or jointly with others, should not exceed 5 ha.

b. The loan amount shall not be used for acquiring any additional land.

- *Purpose:* To finance farmers only for land-based activities and to carryon agricultural activities on the existing land.
- *Amount of Loan*: The maximum amount of the loan will be need based.
- *Rate of Interest*: Interest rate ranges from 2.50 per cent below to 1.50 per cent above BPLR for various short-term limits and from 1.75 per cent below to 2.00 per cent above BPLR for various long-term limits.
- *Repayment*: The loan can be repaid out of the income generated from the agricultural activities or remittances from abroad or by debit to their NRE/NRO/FCNR accounts.

Minor Irrigation

Projects with cumulative command area of less than 2000 ha are called minor irrigation projects:

- *Eligibility*: The beneficiary should have a minimum of 50 cents of land to be brought under irrigation to ensure viability and repayment of loan.
- *Purpose*: Scheme for developing irrigation potential, Minor Irrigation, Installation of Pump set Drip Irrigation etc.
- *Amount of Loan*: As per the project submitted.
- *Rate of Interest*: Interest rate ranges from 1.75 per cent below to 2.00 per cent above BPLR for various limits.
- *Repayment*: The loan shall be repaid within a period of 9 years, in yearly installments.

Farm Mechanisation

Loan for Farm Mechanisation, Purchase of tractors, Power Tillers, etc.

- *Eligibility*:
  a. Tractors with engine capacity up to 35 HP – The applicant should own/ cultivate six acres of perennially irrigated land.
  b. Tractors with engine capacity above 35 HP – The applicant should own/ cultivate eight acres of perennially irrigated land.
  c. Power Tillers – the applicant should own/ cultivate four acres of perennially irrigated land.

- *Purpose*: To purchase tractor/power tillers for agricultural activities.
- *Amount of Loan*: Amount of advance will be the investment cost of tractor/ power tiller and implements less margin @15 per cent.
- *Rate of Interest*: Interest rate ranges from 1.75 per cent below to 2.00 per cent above BPLR for various limits.
- *Repayment*: The period of repayment shall be 9 years for tractors and 7 years for power tillers.

Scheme for Cultivation of Medicinal Plants

- *Eligibility*: All agriculturists are eligible.
- *Purpose*: Scheme for financing cultivation of 22 medicinal plants cultivated extensively and also in great demand in the local as well as foreign market.
- *Amount of Loan*: Depending on the area of cultivation/ project cost
- *Rate of Interest*: Interest rate ranges from 1.75 per cent below to 2.00 per cent above BPLR for various limits.
- *Repayment*: Repayment should coincide with harvesting and marketing or at the time generation of income from the scheme.

Scheme for Cultivation of Vanilla

- *Eligibility*: All agriculturists are eligible.
- *Purpose*: Scheme for financing cultivation of Vanilla, a cash crop, gaining ground in the State of Kerala.
- *Amount of Loan*: Amount of finance will be ₹250000 per hectare for pure crops and ₹210000/- per hectare for intercrop.
- *Rate of Interest*: Normal rate of interest as applicable to ATL.
- *Repayment*: The loan shall be repaid within a period of 7 years, in yearly installments. Farmer's eligible for two years gestation period and interest is repayable on the 3rd and 4th year and the principal from the 5th to 7th year.

SBT Rainwater Harvesting Scheme

- *Eligibility*: Farmers having land holding of 0.50 acre or more are eligible to be considered for finance under this scheme.
- *Purpose*: Scheme envisages construction of low cost tanks for collecting and storing rainwater and using it for irrigation, by siphon arrangement, utilizing gravitation flow or by installing motor pump.

- *Amount of Loan*: Maximum amount of finance will be ₹88000/- per acre. Scheme can be adopted in smaller areas also by reducing the cost proportionately.
- *Rate of Interest:* Interest rate ranges from 1.75 per cent below to 2.00 per cent above BPLR for various limits.
- *Repayment*: Repayment based on the income generated from the crops raised and cropping pattern. The maximum period eligible for repayment is 8 years in annual instalments.

Agriculture Gold Loan

- *Eligibility*: All individual farmers undertaking cultivation or other activities including allied activities are eligible for short-term finance.
- *Purpose*: To meet genuine credit requirements of farming including allied activities, repairing of equipments and consumption needs etc.
- *Amount of Loan*: The eligible loan amount should be assessed based on the area under cultivation, crops(s) raised, scale of finance and not in relation to the value of gold offered as security.
- *Rate of Interest*: Interest rate ranges from 2.50 per cent below to 1.50 per cent above BPLR for various limits. For working capital loans like ACC/KCC/AGL up to ₹ 3 lakh interest at the rate of 7 per cent is extended as per RBI guidelines subject to the periods stipulated by RBI and beyond that normal rate will apply.
- *Repayment*: As applicable to Agri. Cash Credit accounts depending on the duration of crops raised and harvesting period and income generation, subject to a maximum period of 12 months. The account has to be closed at the end of the repayment period.

Scheme for Development/ Strengthening of Agri. Marketing Infrastructure, Grading and Standardization

- *Eligibility*: Scheme shall be available to individuals, groups of farmers/ growers/ consumers, partnership/partnership firms, NGO's, SHG, Companies, Corporations, Cooperatives, Co-marketing Federations, Local Bodies, etc.
- *Purpose*: For development of agricultural marketing operations including strengthening of infrastructure, techniques of preservation, storage, etc.
- *Amount of Loan*: As per the project.
- *Rate of Interest*: BPLR irrespective of credit size.

- *Repayment*: Adequate long-term repayment period according to the project.

Construction/ Renovation/ Expansion of Rural godown

- *Eligibility*: The project for construction of rural godowns can be taken up by Individuals, Farmers, Group of farmers/growers, Partnership/Proprietary firms, NGOs, SHGs, Companies, Corporations, Co-operatives, Federations, Agricultural Produce Marketing Committees, Marketing Boards and Agro Processing Corporations.
- *Purpose*: To create scientific storage capacity with allied facilities in rural areas to meet the requirements of farmers for storing farm produce, processed farm produce and agricultural inputs.
- *Amount of Loan*: As per the project.
- *Rate of Interest*: As applicable to advances under SIB/ CandI segments will be charged.
- *Repayment*: Adequate long-term repayment period, not less than 5 years including a grace period of one year.

finance for agriculture and food processing

Banks: The Food Processing sector has access to credit from Commercial Banks-Indian and Foreign, Cooperative Banks and the Regional Rural Banks. The formal institutional sector assists the food processing companies through long term loans for capital investments and short term loans for working capital. The total lending to the food processing sector is clubbed with lending to the agriculture sector and separate details for financing to the food processing sector are not available.

- National Bank for Agricultural and Rural Development (NABARD): NABARD offers refinance facilities for food processing, agri infrastructure, developmental assistance for RRBs, DCCBs, SSI and SHG linkages, and assists in infrastructure development and research.
- Small Industries Development Bank of India (SIDBI):SIDBI has been involved in assisting the entire SSI Sector including tiny, village and cottage industries through suitable schemes tailored to address the funding requirements for expansion, diversification, modernisation and rehabilitation.
- Export Import Bank (EXIM Bank): Exim Bank assists in financing and facilitation of foreign trade. The Bank also offers financial support to companies engaged in exports.

- National Cooperative Development Corporation (NCDC): NCDC assists in promotion, planning and financing the agricultural supply chain from production, processing, storage and trade of agricultural produce and food products. NCDC also provides assistance for marketing of certain notified commodities *e.g.* fertilizers, pesticides, agricultural machinery etc.
- Ministries/Government bodies
- Ministry of Food Processing Industries (MFPI): MFPI is the Nodal agency for development of the processed food sector in the country MFPI's financial schemes include schemes for technology upgradation, Human resource development, Quality testing, R&D,TQM, backward and forward integration, development of infrastructure including food parks.
- Agricultural and Processed Foods Products Export Development Authority (APEDA): APEDA facilitates market linkages between Indian producers, manufacturers and the international market. APEDA provides financial assistance for market development, infrastructure development and development of quality enhancing facilities.
- Ministry of Agriculture (MoA), Government of India: The Ministry of Agriculture under various schemes, provides financial assistance for development of specific crops for investment in seeds, irrigation, farm implements, inputs, infrastructure and training.
- National Horticultural Board (NHB):NHB promotes integrated development in horticulture, assists in development of post harvest management infrastructure, promotes production and processing of fruits and vegetables, strengthening of market information systems and assists in R&D programmes in cultivation and processing. NHB's financial schemes are directed towards commercial horticulture and infrastructure related to post harvest techniques.

Financial assistance from these organizations are in the form of grants, back-ended subsidies, soft loans, refinance etc., with most of the schemes directed to specific sub-sectors of the agri/food processing industry. In addition, some State Governments offer financial schemes to companies in the sector.

The key issues with Government schemes are highlighted below:

- Overlap among schemes: Various Government agencies have separate schemes which aim to achieve the same objectives. Thus, an applicant can potentially approach different agencies and obtain assistance from more than one agency. Some of the examples of overlapping schemes are as follows:

- Several areas addressed inadequately: There are a host of areas which are not addressed by these schemes such as financing of freezer cabinets for retail outlets, financing of bulkcoolers for milk, financing of alternate markets/mandis, financing of vending machines for tea/coffee/beverages etc.
- Creation of excess capacity: There are instances where assistance has been provided to set up cold storages in locations which have excess storage capacity. This has lead to an unviable situation for all cold storage operators in the region. It is essential for the nodal agencies to achieve development of the food processing sector in a sustainable manner.
- Inadequate financial assistance^ most instances, there is a cap on total financial assistance provided at INR 5-5.5 mn. This leads to fragmentation in capacity creation, and often, a situation of overcapacity. At the same time, it does not allow players to scale up their operations.
- Inadequate Monitoring of progress of schemes: It is important to track progress of the projects funded by the Government. Periodic appraisal of the progress will ensure that funds are appropriately utilized. This will also enable assessment of success and lacunae of these schemes and record the multiplier effect of the financial support provided.
- Most of the schemes are back-ended which leads to a funds crunch during implementation.
- Institutions providing subsidies do not undertake their own assessment and rely on the appraisal report of the bank which provides loans to the project.
- There are significant time lags from the date of application for financial assistance, to release of funds, and may affect the project schedule, and lead to cost overruns
- These schemes do not address working capital requirements, which due to the nature of the industry are significant and often higher in terms of quantum than capital investment requirements
- Applicants claim that they are often unaware that their applications have been rejected on account of lack of communication from the funding institution. Further, applicants are unaware of the reasons for rejection.

Sources of finance for Agriculture

At present nearly 60 per cent of rural credit is used for agriculture and allied activities, 10 per cent for non-farm activities and the balance 30 per cent for funding household consumption expenditure. Nearly 40-45 per cent of rural credit needs are catered to by formal credit institutions and the balance by the informal sector including commission agents, input suppliers, traders, Self Help Groups (Joint Liability Groups), processing industries and professional money lenders.

The Indian commercial banks are mandated by the Reserve Bank of India (RBI), to undertake directed lending for agriculture and rural development. Commercial banks are required to achieve priority sector lending of 40 per cent of net bank credit of which 18 per cent should be for agriculture.

Further, sub-targets are also specified *i.e.*, Direct Agriculture (credit for direct farmer benefit, Target: Minimum of 13.5 per cent of the net bank credit) and Indirect Agriculture (Target: Maximum of 4.5 per cent of the net bank credit). In FY 03, total agricultural advances outstanding under priority sector lending, by Public Sector Banks stood at INR 735 billion, on a total net bank credit portfolio of INR 4779 billion. As compared to this, agricultural advances by Private Sector Banks stood at INR 119 billion, on a total net bank credit portfolio of INR 718 billion. The following Exhibit provides details of direct and indirect priority sector credit amongst Public and Private sector banks during FY03.

In case of a shortfall in lending to agriculture, banks are required to invest in Rural Infrastructure Development Fund (RIDF) deposits, which offers interest rates linked to the bank's performance in lending to agriculture.

These rates are inversely proportional to the shortfall in agricultural lending. The interest rate on RIDF is provided in slabs, based on the achievement of the priority sector lending target, and can be as low as 4 per cent to 5 per cent.Therefore, it is imperative to achieve the agriculture target and more specifically the Direct Agriculture target as it accounts for 75 per cent of the total.

Established in 1996, RIDF is being utilized for making investments in rural infrastructure projects. Nine tranches of RIDF have been established with an aggregate corpus of INR 340 billion, with the ninth tranche of INR 55 billion. Cumulative sanctions and disbursements under various tranches of RIDF stood at INR 334 billion and INR 193 billion respectively, at the end of February 2004.The RIDF funds are used to finance minor irrigation projects, shallow wells, drinking water facilities, roads, bridges and storage infrastructure.

Financing to Agriculture and Food Processing

In order to improve the funding from the institutional set-up to agriculture and food processing, the following policy changes are required:

Restructuring of Government Schemes

- In capital intensive projects, the interest can be capitalized and funded, so that the burden of high cost of interest is not faced by the company in the formative years
- The agencies could structure schemes in the form of capital subsidies, which in turn could be leveraged by the companies to access bank finance. Release of funds should be proportionate to the equity brought in by the promoters.
- In sectors, which are working capital intensive, the funding institutions should consider offering soft loans instead of grants. The soft loan would ensure that the companies undertake operations to service the interest payable to the institution.
- Market analysis prior to providing funding-Detailed market analysis of demand-supply trends, costs, competitive position should be undertaken by the funding institution/by a neutral third party which has the required industry expertise, to ascertain the financial viability of the project
- In order to address the issues related to food parks, the recommendations are as follows:
- It is recommended that food parks be located in proximity to raw material sources to maximize the potential available locally. This will ensure better utilization of common facilities.
- The quantum of financial assistance to a food park needs to be decided on a case-specific basis, depending on the infrastructure requirements across locations. The financial appraisal of any food park scheme should be evaluated on the basis of potential for tie-ups with off takers and sufficient availability of raw materials.
- The definition of common facilities could be expanded to include roads, drainage and solid waste management facilities etc.
- It is critical to have a monitoring mechanism to facilitate and review progress of implementation. The State Departments for food processing can play an active role in this process.

Redefine role of Regional Rural Banks (RRBs)

- The RRB Act needs to be amended to permit RRBs to close branches and allow merger/amalgamation of RRBs among themselves or with the sponsor banks to take advantage of operational synergies. The capital structure of the RRBs also needs to be revised since state governments have been unwilling to inject their share of capital in past recapitalization exercises. The government may permit sponsor banks to increase their stake in the RRBs or bring in new investors. With this reorientation, RRBs can be effectively utilized by banks and Financial Institutions as channels for rural lending. B. Amendments in Priority sector Lending

- Definition of priority sector: In a recent circular, Reserve Bank of India has imposed a cap on investment in plant and machinery upto INR 50 million for food processing industries for classification under the priority sector. As also detailed in subsequent sectoral sections, for Indian food companies to compete effectively, the policy environment needs to address all bottlenecks hindering scale of operations, and availability of financing features high on this list.
- All food processing units should be classified as priority sector irrespective of size. Relaxation of the cap on investment for classification as priority sector would provide the required fillip for flow of credit to the agribusiness sector and would also have a beneficial impact on farmers.
- Further, the scope of priority sector lending should be increased to include all primary processors of agriculture produce under 'priority sector' category, including sugar manufacturing units, rice mills, dal mills etc.
- Farmer financing against warehousing of agricultural produce: Priority sector norms stipulate a cap on advances against warehouse receipts of produce upto INR 0.5 million. The norm also stipulates the advance to be exclusive to farmers availing a crop loan with the same Bank This cap on the quantum of borrowing should be removed. Further, the advance should be delinked from the crop loan. This will encourage farmers to store their produce during the peak season and sell it during the lean period, to obtain better prices for their produce. The current system forces the farmer to sell all his produce immediately after harvest to traders/commission agents, who store the produce and profit due to price fluctuations.
- Financing for high value crops: Cultivation of crops such as grapes requires high upfront investments, estimated at INR 0.8 1.0 million per acre for establishing of vineyards. The cap on farmer loans for a maximum amount of INR 0.5 million per farmer restricts bank finance. This cap on farmer loans should be made category/crop specific, taking into account the respective cultivation and investment expenses.

- At present, banks typically provide financial assistance through loans for a period of 3-4 years. However, infrastructure such as temperature controlled storage have a longer gestation period. There should be a specific focus on encouraging banks to provide long term loans for a period of 5-7 years, which should be classified as direct agriculture lending.
- Lending to agriculture is usually dictated by the scale of finance method of assessment of the credit requirement. This system is outdated as the norms are not based on current costs of cultivation. Banks need to assess credit needs of farmers based on current cultivation costs.
- As mentioned earlier, priority sector lending records the outstanding credit in the books of the banks on the last working Friday of March every year. RBI and NABARD should monitor banks' outstanding portfolio at least twice in a year, so that banks undertake mandatory lending to the agriculture sector during the kharif season as well.

Promote Warehouse Receipt-based Financing

Warehouse receipt finance is a form of secured lending to owners of commodities, (farmers, traders, processors) which are stored in a warehouse, with the commodities having been assigned to a bank through warehouse receipts.

On the back of the warehouse receipts being housed with the bank, the latter has the security of goods until they have been sold and the proceeds collected. Warehouse receipts are a proven tool for financing, utilizing the stored goods as collateral for loans. The system has been working successfully from the beginning of the twentieth century in several countries such as Brazil, Indonesia, Singapore and Argentina. In addition to addressing the needs of food processors, as mentioned earlier, farmers can also benefit because it enables them to store produce and sell when prices are favourable.

Some of the advantages of a warehouse receipt system are:

- It provides a choice to primary producers who can decide to sell immediately after harvest or to store in a licensed warehouse and to apply for short-term credit. The farmer can decide to sell his crop later in the year, when prices are typically higher than at harvest time.
- It enables primary processors to purchase raw materials at favourable prices and store until required for processing

- The system leads to a reduction of post-harvest losses as the produce is stored under appropriate conditions in licensed warehouses.
- It translates into lower risks for the financier, as the collateral for the loan is a liquid asset.
- This can lend greater depth to operations of Commodity Exchanges, by increasing the number of transactions without physical movement of goods, through endorsement of the warehouse receipts

The Government can play an important role in promoting warehouse receipt-based financing through the following measures:

- Amend the Negotiable instruments act to introduce negotiability of warehouse receipts
- Creation of an appropriate legal environment to ensure ease in enforcement of security
- Facilitate development of a network of warehouses at appropriate locations
- Facilitate direct processor farmer linkages (rather than via the mandi) to enable warehouse receipt based financing
- Dematerialisation of warehouse receipts to enable electronic trading

Infrastructure and Project finance

Infrastructure development can play an important role in improving efficiency across the supply chain and hence any financial assistance for rural infrastructure development including rural roads, markets, silos etc. should be classified as priority sector. It is recommended that the Government undertake the following measures to boost infrastructure creation for the agriculture and food sector-

- Creating an enabling legal and regulatory frame work for development of infrastructure
- Providing seed capital and leveraging the same for development of agri-infrastructure
- Monitoring and facilitating private sector participation
- Risk sharing with banks in financing agricultural infrastructure projects.

Public Private partnerships are key to development of agricultural infrastructure projects. The linkages between various players, which includes the users, the supply chain participants and operators are crucial. The projects can be structured based on the BOT (and its variant) model which ensures appropriate risk allocation between various players, including the Government.

Some of the project structures which can be adopted are as follows:

- User Charge-based project structure: A Special Purpose Vehicle (SPV), funded by the sponsors and the financier is constituted. The Users, under a waterfall agreement pay on usage of the facility. These user charges service the debt.
- Rated lease-based project structures: The Sponsors assign the assets to a Special Purpose Vehicle (SPV), financed through a mix of debt and equity (from financial investors).The SPV executes lease agreements with the lessee, who in turn pays lease rentals, which are used for regular debt servicing.
- Receivables-based structure: This type of structure can be used for construction of rural roads or market yards, wherein the project is financed based on future receivables from a determined and realizable set of cash flows.

Financing-food Processing Sector

The food processing sector comprises a large number of small and medium sized companies, a significant proportion of which have stand alone operations, with no control over the raw material base and reliant on other organizations to undertake marketing/further processing of their products. Banks and financial Institutions adopt the same risk models relevant to the manufacturing sector, for assessing food processing companies. Interest charges for food processing companies are high, on account of the high risk perception associated with the nature of their operations.

Food Processors' Perspective

With raw material availability often being seasonal, and/or concentration in demand, inventory holdings are high and thus the working capital finance issued through normal Maximum Permissible Bank Finance method does not address the funding requirements adequately. The rate of interest for working capital assistance is high. Most food processing companies are not able to access adequate working capital at reasonable rates. This affects the raw material procurement and capacity utilization through out the year. Unavailability of large working capital facilities, or facilities without a high seasonal drawal, restricts players in purchasing agri produce in large volumes, when prices are favourable.

Banker's Perspective

- Lack of reliable information on demand-supply, price trends, raw material supplies etc is a significant constraint for banks, impacting credit assessment and monitoring

- High operational/transaction cost in servicing small companies
- Lack of backward linkages for assured access to raw material
- Lack of forward linkages of stand-alone food processing units for marketing/distribution

Owing to high cost of market development, the profitability of food processing companies is under pressure in the initial years of their operation. This impacts the risk rating and leads to high cost of borrowing. Even temporary adversities in market conditions can lead to defaults and eventually the loan can become a non-performing asset for banks.

5

Fibre Crops

Jute

Jute, a bast fibre, is obtained from one of the most important cash crops of eastern India Exported as manufactured goods and as raw fibre, it earns foreign exchange around 3,000 million rupees. The crop is grown in West Bengal, Assam, northern Bihar, south-eastern Orissa, Tripura and eastern Uttar Pradesh on 0.80 million hectares annually and the average production of dry fibre is 6.5 million bales.

Currently, about 82 per cent of the total annual production is used by jute-mills in India, 8 per cent is retained by cultivators and the remainder is exported to Europe and the USA.

The fibre is chiefly used for manufacturing hessian, sacking and carpet backing. It is used for storing and transporting grains, pulses, spices sugar, cement, fertilizer, minerals, cotton and wool all the world over. Jute is also used for making mats, tarpaulins, ropes and twines. Woollenized jute is used for manufacturing cheaper rugs, and mixed with cotton it is used for producing decorative cloth, curtains and upholstery.

The jute sticks are largely used as fuel and also for making gunpowder charcoal. Since recently, the paper industry has been using it as a raw material for coarser paper. Resin-bonded pressed jute sticks make durable hard board.

Climate

Jute requires a warm and humid climate, with temperature fluctuating between 24 °C and 37 °C, the optimum being around 34 °C. The permissible diurnal variation in relative humidity favourable to growth is between 57 and 97 per cent. Incessant rain water-logged conditions are difinitely harmful. Whereas the capsularis varieties can stand water logging to some extent. In the seedling stage, water-logging is not tolerated by both the species.

Areas receiving early showers in late February or early March, followed by a relatively short spell, are suitable fore sowing the capsularis varieties. Olitorius varieties, until the introduction of 'JRO 878' and 'JRO 7835' have been sown with showers in April or even in May. The monsoon sustains the subsequent growth. Rain below 1,000 mm is not helpful. Alternate sunshine and rainy days are most conducive to growth. The dry spell preceeding the break of the monsoon enables the essential weeding and thinning operations to be carried out.

Soil

The new grey alluvial soil of good depth, receiving silt from the annual floods, is nature's best gift, but jute is widely grown in sandy loams and clay loams, with varying soil-management practices. Sandy soils and heavy clays are unsuitable. Soils with a low pH give a poor crop, the optimum pH being around 6.4.

Cultivation

Jute requires a clean, clod-free field with fine tilth. The land is, therefore, ploughed, cross-ploughed, and planked several times. All weeds are thoroughly removed.

Sowing

The sowing of capsularis varieties starts in late February in low-lying areas that retain moisture of the previous flood or monsoon.

Sowing in midlands and highlands starts with showers in March or April and continues till early June in the western part of the jute belt. For broadcast sowing, the seed-rates are 10 and 7 kg per hectare for capsularis and olitorius varieties, the germinability of seed being 80 per cent or above. The high seed-rates ensure an even emergence of seedlings, which are thinned out in broadcast plots to a spacing of 10 cm. This operation is preferably carried out in two instalments, once when the seedlings are about 10 cm and subsequently when they are about 15 cm tall. While thinning, care is taken to remove all weeds with a hand-hoe. These operations account for 30-40 per cent of the cost of cultivation.

Row-cropping has proved advantageous and a single row seed-drill is used to sow the capsularis varieties 30 cm apart and the olitorius varieties 20 cm apart. The plant-to-plant spacing is manually adjusted between 5 and 7.5 cm. Sowing is always done shallow, *i.e.* not exceeding 3-4 cm.

Cropping Pattern under Irrigation

In jute-paddy potato, jute-paddy-wheat, (jute-moong)-paddy-potato, cowpea-jute-potato, and jute-paddy-berseem rotations, jute may need one pre-sowing irrigation and one sustaining irrigation cropping pattern under rainfed conditions. Jute-paddy-mustard, jute-paddy-pulses or jute-paddy rotations are possible under this cropping pattern.

Nutrition and Manuring

The gross uptake of N, P, K, Ca and Mg by the capsularis variety 'JRC 212' is 84, 16, 147, 84 and 29 kg respectively and by the olitorius variety 'JR0 632' is Ill, 28, 164, 124 and 25 kg per ha, respectively. The resulting fresh crop weighs 40 to 50 tonnes, yielding 20 to 25 quintals of dry fibre, which is around 5 per cent of the fresh weight of the crop.

Acid soil require amendment with 3-7 tonnes of lime in 3 years. Compost or farmyard manure at the rate of 4-7 tonnes per hectare promotes proper growth, specially when fertilizers are not available. P and K are applied as basal nutrients, whereas nitrogen is better top-dressed in two instalments-N (40-80 kg per ha for capsularis varieties, 20-60 kg per ha for olitorius varieties)-P (half of the quantity of N) and K (quantity equal to N). If the jute crop is preceded by potato, no fertilization is recommended; if it is preceded by paddy or wheat, P and K are not so much required. Nitrogen is the most important nutrient required. The ash of water-hyacinth is a rich source of potash. The application of magnesium on the North Bank of Brahmaputra and in northern Bengal produced good results.

In case of short supply of nitrogen, foliar feeding with nitrogen at the rate of 15 kg of urea per ha at 10 per cent concentration, with a low-volume power sprayer in 2 instalments between 35 and 60 days' age of the crop proved useful. Capsularis varieties respond better up to 30 kg of urea.

Interculture

Weeding, specially in the early stage, is a must. For a broadcast crop, the pre-sowing application of 2,2,3,3 tetrafluorpropionate of sodium proved to be beneficial. A post-emergence application of mono-sodium methanarsonate with or without 2,2 dichloropropionate, as a directed spray, proved to be beneficial in row-cropped fields. Yields, with two hand-weedings, however, are significantly superior and constitute the prevailing standard practice. Smothering the weeds between rows with a wheel-hoe saves labour and helps to mulch the soil.

Major Pests

The female of jute apion (Apion corchori Marshall) lays an egg at a node where a knot is formed; it makes one puncture or several on the same plant. The jute semilooper (Anomis sabulifera Guen.) eats tender leaves and is widespread; it usually comes in three waves between late June and August. The vehow mite (Hemitarsonemus Iatus Banks) sucks the sap from young apical leaves which curl, and the growth of the plant is arrested. The hairy caterpillar (Diacrisia obliqua Walker), when young, is gregarious on a single leaf and later scaterrs out and feeds on the foliage.

The root-knot nematode (Meloidogyne incognita Kofoid and White) attacks the capsularis varieties more than the olitorivg varieties and induces chlorotic symptoms and arrests growth. What rovides foci for root infection bataticola (Taub.) Butler. hytrypes achatinus Stoll is a menace to seedlings which are cut. The pest is mostly confined to Assam.

Minor Pests

The indigo caterpillar (Lapltygma exigua Habn.), the leaf-miner (Trachys dasi Thery.), the red mite (Olygonychus coyfeae Neither), the jute stem- girdler (Nupserha bicolor postbrunnea, and the leaf-eating weevil (Myllourus discolor Boheman) are not of commercial importance.

Control Measures

DDT and Gammexane, with a limited efficacy in the field, are unprofitable. Ethyl parathion and Endrin are highly efficacious against all major pests. But because of their hazardous nature, they have been withdrawn. Endosulfan, Leptophos, Phosalone, Fenitrothion, Chlorfenvinphos are effective to different extents against the major pests; Monocrotophos and Tamaron on semilooper; lime sulphur on yellow mite and Heptachlor and Aldrin dust on crickets and Dicofol on red mite, are effective.

Timely sowing and clean cultivation guard against afion; hand-picking of the leaves with gregarious clusters young hairy caterpillars controls the pests.

Diseases

The important diseases are:

- Seedling blight, stem-rot, collar-rot, and root-rot caused by Macrophomina phaseoli (Maubl.) Ashby, both seed-and soil-borne, secondary spread through air-borne pycnospores. The predisposing factors are low pH of the soil, the lack of adequate potash water-logging, high temperature and excess nitrogen. Seed treatment prevents seedling blight, whereas soil amendment with lime and potash retards its incidence. Jute-paddy and jute-paddy-wheat rotations are useful in reducing the incidence of the disease.
- Hooghly-wilt, caused by a microbial complex, including Macrophomina phaseoli, Fusarium solani. Pseudomonas solanacearum, is confined to the jute potato fields only; control through rotation with paddy and legumes is possible.
- Colletotrichum gloeosporioides causes anthracnose in Assam.
- Leaf mosaic is caused by a pollen-transmitted seed-borne virus,

also by white fly Bemisia sp. in Assam, specially in capsularis varieties.

- Physoderma stem gall on the seedlings of olitorious varieties occurs in Uttar Pradesh and Tripura.
- Soft rot of stem Pellicularia rolfsii (Sace.); the disease can be controlled with deep ploughing, and clean cultivation.
- Nematodes provide foci for infection by Macro phomina phaseoli in sick soils.

Harvesting

Jute may be harvested any time between 120 and 150 days. Early harvesting gives finer fibre of good quality, whereas late harvesting gives a larger yield but a coarser fibre. The compromise between quality and quantity is found in harvesting at the early pod stage or around 135 days of cropping. The other consideration for early harvesting is to accommodate paddy-transplanting in accordance with the cropping pattern chosen.

Harvesting is done by cutting the plants at or close to the ground level. In flooded land, plants are uprooted. The harvested plants are left in field for 2 to 3 days for the leaves to shed. Next, the plants are tied into bundles, 20-25 cm in diameter, and the branching tops are lopped off to rot in the field.

Retting

The bundles are kept standing in water, 30 cm deep, and later placed side by side in retting water, usually in 2 to 3 layers and are tied together. They are covered with water-hyacinth or any weed that does not release tannin and iron. The float is then weighed down with seasoned logs or with concrete blocks or are kept submerged (at least 10 cm below the surface of the water) with bamboo-crating. Clods of earth used as a covering material or as weighing agent produces dark (shyamla) fibre of low value.

Gently flowing, fairly deep, clear and soft water is ideal for retting. The optimum temperature is around 34 °C; ditches, tanks and pools are also used for retting. Incomplete submergence produces 'croppy' fibre of extremely low value. Most of the defects in fibre are due to faulty retting. Over-retting results in 'dazed' weak fibre.

Retting is a microbiological process and, therefore, the end-point is determined by inspecting a few plants each day from the tenth day onwards. If fibre slips out easily from the wood on pressure from the thumb and fingers, retting is considered complete.

Extraction

The fibre, when extracted separately from each reed (stem) with fingers; is sleek, clean and free from entanglement. By the beat-break-jerk method, ten or twelve reeds are taken at a time; their stiffer root-ends are beaten with a mallet to loosen the fibre. The bundle is then broken in the middle and the fibre is loosened. By gripping this loosened fibre in the middle, the broken bundle is jerked in water so that the sticks slip off. The fibre is then washed in clean water, wrung and. eventually spread to dry, preferably in shade or mild sun. This second method often leaves the broken sticks and makes fibre somewhat entangled, resulting in 'sticky', fibre.

The extraction of fibre from the green stem with a machine, followed by a short-period retting has proved to be successful. Adaptive research continues at the Jute Agricultural Research Institute.

Yield

The national average is 13 quintals, per hectare. In general, an olitorius variety gives better yield than a capsularis variety, and the former is a more efficient user of nitrogen. Under the National Demonstration Programme, the average yield has been improved to 27 quintals in the case of the olitorius and to 20 quintals in the case of the capsularis varieties, the maximum potential being 40 quintals and 37 quintals, respectively.

A strong positive correlation exists between the plant height (PH) and the yield, between the base diameter of the stem (BD) and the yield, also between PH and BD. Hence, taller and stouter plants give higher yield. However, if the yield from a smaller number of stout plants is equal to that from a greater number of thinner plants, the quality in the case of the latter is always better and, there-fore, preferred. The leaves of both the species are used for culinary purposes, olitorius being sweet and capsularis being bitter, but the latter is considered to be good for the stomach.

Varieties

The capsularis varieties cover 60 per cent of the jute area, whereas the olitorius varieties cover 40 per cent of the area, mostly in the southern West Bengal.

Released Capsularis Varieties

- 'JRC 321' (Sonali) has a light coppery-red stem. It is suitable for low-lying areas for early sowing from late February to mid-March. It comes to 50 per cent flowering in about 125 days (variable); it is fast-growing and has good-quality fibre. It yields 20-25 g per hectare in northern West Bengal and Bihar.

- 'JRC 212' (Sabuj sona). This variety has a full-green stem. The sowing time is from late March to mid-April. It is suitable for medium and highlands where inundation does not occur. It is particularly good for Assam, Bihar, northern West Bengal, Uttar Pradesh and Orissa. With 60 kg N per ha, it yields around 30 q per ha.
- 'JRC 7447' (Shyamali). This variety has a full-green stem, and is an X-ray derivative of 'JRC 212'. It responds better to higher doses of N above 60 kg per ha; it is suitable for Orissa, Nadia, Murshidabad and 24-Parganas.
- 'D 154'. It is an old standard variety, originally selected in 1918 from 'Kakya Bombai'. It is a hardy variety and can adapt itself to all conditions. It yields 20-25 q per ha.
- 'JBC 1108'. A pre-release variety. It yields better than 'D 154', equalling 'JRC 212' at places, and is fairly resistant to stem-rot and root-rot.
- 'JRO 632' (Baisakhi tossa). This variety has a full-green stem; its sowing time is from 15 April onwards on midlands and highlands. It takes 130-140 days to flower; its seed colour is steel grey; its yield potential is 40 q per ha; its yield averages 32 q per ha with 40:20:40 "NPK plus farmyard manure. Its fibre is of good quality.
- 'JRO 878' (Chaitali tossa). The stem of this variety is red-pigmented. Its sowing time is the second week of March to mid-April on highlands. It is non-lodging and has non-shattering pods. It does not flower prematurely. It is suitable for West Bengal and Orissa. Its Yield Potential is 38 q and the average yield is 32 q per ha, with 40 N: 20 P: 40 K plus farmyard manure.
- 'JRO 7835' (Basudev). The stem of this variety is green. It is sown in early March to early May. It is non-lodging and bears non-shattering pods (It does not flower prematurely. It is suitable for all areas with highlands). It is high-yielding and produces around 34 q per ha with 40: 20. 40 NPK plus farmyard manure.
- 'JBO 620'. The stem of this variety is red-pigmented. It yields fibre of superior quality. It yields 3-4 quintals less than 'JRO 632' and, therefore, has gone out of cultivation.
- 'C.G.' (Chinsurah green) is a Poor Yielder and has been withdrawn. The pre-release variety 'JRO 524' (Navin), yields as high as, or more than 'JRO 7835', with added qualities of fine fibre and a stem that rets quickly.

Seed Selection

Since jute is harvested much before seed formation, special arrangements to raise the seed-jute Crop. (The Cultivators often leave about 4 per cent of the fibre crop for seed production.) The seed crop is sown later than the fibre crop for seed production. Some interculture is necessary to keep down weeds. Unlike the fibre crop, the seed crop is spaced wider, so that branching is promoted.

The induction of branching by clipping the tips at the right time and with right stage of growth, with assured nutrition for the development of branches, has been successfully practised. This practice, however, would not be resorted in the areas with low rainfall and where land is not very fertile. Natural cross-pollination being low genetic mixture is avoided by alternating capsularis and olitorius seed plots. The crop requires much less nitrogen for proper growth. The average yield of seed is 3.8 to 5.0 q per ha in the case of capsularis varieties and 1.7 to 2.5 q per ha in the case of 'JRO 878' and 'JRO 7835' and around 3 q per ha in the case of 'JRO 632'. After harvesting the crop the seed is dried in the sun for about 4 days and stored in gunny bags lined with plastic cloth and kept in dry godowns, avoiding contact with the floor. Corynespora casseicola and powdery mildew attack pods and capsules. These pathogens are controlled with copper fungicides. Die-back (Diplodia corchori) on the seed crop is rare, but serious.

Defects in the Fibre

The defects in the fibre are its being barky and croppy (incomplete submerging) or incomplete retting; specky (apion damage); leafy (incomplete retting or washing); and sticky (defective extraction). The fibre becomes dirty by using clods of earth and muddy retting water and careless washing. The plants become mossy with adventitious roots emerging from the stems. The fibre becomes shyamla as a result of iron-tanning reaction. 'Roots are, caused by more than one factor but the main one is 'incomplete separation of fibre at the base of the stem during retting.'

Quality

The quality of the jute fibre among varieties differs on the basis of anatomical conditions and is genetically controlled. Coarser and light-body fibre is obtained from sandy soils whereas clay-loam soils with silt give fibre of superior quality. Climate and the nutrition pattern also affects the fibre. But the most important single factor is 'retting' which, if faulty, mars the positive contributions of the variety, soil, climate etc. Under-retting gives coarse and over retting dazed and weak fibre.

Marketing

Jute Corporation of India, Calcutta, has the objective of stabilizing the price line, both in the interest of the farmers and the consuming mills.

Grading

Grading system for white jute (capsularis) has 8 classes, *viz.*, WI to W8, on the basis of length, strength, fineness and lustre, and freedom from enlargement and 'roots' (rejectable basal portion). For Tossa jute (olitorius) there are 8 classes, *viz.* TDI to TDS.

Mesta

The fibre of mesta is obtained from stem of Hibiscus sabdariffa variety altissima (n = 36) and H. cannaunns (n = 18), family Malvaceae. JIS and RC denote varieties of the two species.

Production

The area under mesta has been increasing since 1952 and is around 0.24 million hectares, of which more than 80 per cent is under HS types. The annual average production is around 1.08 million bales (1 bale = 181 kg). Within the jutebelt, the yield of mesta is high in West Bengal, Bihar, Assam and Tripura. Outside side the jute belt, the highest acreage is in Andhra Pradesh and the areas assuming importance are Dandakaranya, Orissa, eastern Madhya Pradesh, Maharashtra, Karnataka and Uttar Pradesh. Other states grow these types over small areas, mostly for home consumption. Yield in the drier areas is less than that in humid ones. At present, the cultivators, in general, grow mesta in poorer soils without much care and profits are marginal. The average yield in different states is between 1.4 and 2 bales. There is great scope of improvement of yield with improved varieties and techniques.

Climate

A warm and humid climate suits both HS and HC, but both grow in drier rainfed areas, the latter being more drought-resistant. In areas with 500-900 mm of rainfall, HC suits better by virtue of its shorter duration and faster growth. Neither of them can stand prolonged water-logging. Both are kharif crops and are sown in April-June with the first showers of the monsoon. Heavy, continuous rains and low temperature are harmful.

Soil

Whereas the rich loams give the highest yield, both grow on a variety of soils, including new and old alluvium; but acid soils are not suitable without amendment. With a high pH (above 7.0) of the soil, chlorosis appears in HS mesta. Both are unfit for low-lying areas, subject to inundation.

Cultivation

Soil is prepared in much the same way as in the case of jute. HS when sown broadcast, requires 12 to 17 kg of seed per hectare. Seedlings, when 10-12 cm tall, are manually thinned in two instalments to a spacing of 15-18 between plants. When raised in rows, seed is drilled 30 cm apart, the seed requirement being 10-12.5 kg per hectare. Seedlings are manually thinned to 12-15 cm apart in the rows. A minor adjustment in spacing is permissible on the basis of the fertility of the soil, rainfall and the nature of the soil. Variety 'HS 4288' (stem with bristles) and 'HS 7910' (stem without bristles) is sown best in mid-April to mid-May, roughly north of 22 °N, early sowing giving better yield. Variety 'AMV 1' is suitable for areas south of 22 °N and can be sown closer to shade the surface of the soil to conserve moisture (22.5 cm × 10 cm) and can be sown till June as in Andhra Pradesh. Dry sowing is possible in anticipation of rain-fall, specially where drizzles do not precede the actual monsoon rains. Variety 'HC 583' requires 17-22 kg of seed per hectare for broadcast sowing and 15-17 kg for line-sowing (30 cm × 12-15 cm). April sowings are best for yield and quality.

Both HS and HC compete well with usual weeds and, unlike jute, can permit delayed weeding. However, two weedings and thinnings are extremely useful, specially when done timely. The use of a wheel-hoe in the row crop, at least twice, within 45 days of sowing leads to better crop.

Nutrition

'HS 4288', yielding 21 quintals of fibre per hectare, lifts 106, 31, 148, 153 and 34 kg per hectare of N, P, K, Ca, and Mg respectively. The corresponding figures for 'HO 867' are 94, 26, 120, 137 and 29. US types respond well to the application of nitrogen from 25 to 60 kg per hectare, depending on the soil type, the method of application and, the rainfall. In drier areas, *e.g.* Andhra Pradesh, a basal dose of 20 kg of N has been found consistently giving 25 per cent higher yield. In areas having a better rainfall pattern, the quantity of nitrogen can be increased to 40 kg per hectare, and the application may be one-third of the basal dose and the remaining may be, top-dressed in one or two instalments between 45 and 70 days of sowing. Although the application to soil ensures best results, nitrogen as urea can be fed through foliage at the rate of 15 kg of N per hectare. Five kilogrammes of N in 100 litres of water is sprayed with a low-volume power sprayer. A single spray is effective between 30 and 75 days from the time of germination.

A high pH of soil leads to chlorosis in HS varieties because of iron deficiency; such soils are not to be used for growing mesta.

Harvesting

'HS 4288' and 'AMV 1' are harvested when 50 per cent of the plant population is in flower; delayed harvesting gives more fibre, but of coarser quality. 'HC 583' is harvested usually a month or more earlier. In Andhra Pradesh, plants are harvested by uprooting, which practice is not recommended; in the jute belt, they are cut close to the ground, as in the case of jute. All US types are of longer duration (180-210), whereas HC flowers in about 150 days after sowing.

Retting

Retting is done in the same manner as in the case of jute, but the low temperature and the paucity of water or both pose a problem. The harvested stems of both HS and HC can be stored under dry conditions for retting next season during the middle of the monsoon. The resulting HC fibre is better than that of HS. Retting in the current season, if possible, is always preferable.

Pests and Diseases

HC mesta once dominated over the HS type in Andhra Pradesh, but it yielded ground to the latter in recent years, since pest and diseases took a heavy toll of the nondescript local types. 'HC 583' evolved at the Jute Agricultural Research Institute can now defend itself against most of the diseases and Pests.

The important pests are as follows:

- Agrilub acuius Thumb is a stem-borer, reaching the stem from the soil. Endrin EC 20 (0.04) is effective, but is to be replaced by Methyl Parathion. The fibre from damaged plants is useless.
- Root-knot nematode, causes dwarfing and chlorosis in extreme cases. Its control, with chemicals is costly; therefore rotation is advocated.

The important diseases are:

- Anthraonose caused by Colletotriehum hibisci Poll. (syn. Volutella sp.) has a withering efflect. It is prevented with Copper oxychloride (50% Cu) at the rate of 3 kg per hectare.
- Tip-rot caused by PMma sp. (syn. Trichosphaeria sp.) responds to copper fungicide;.

- Root-rot is caused by Rhizootonia bataticola alone or in combination with Fusarium oxysporum. Seed treatment with organo-merouric compounds or with TWM or Captan is effective.
- Eye-rot of the stem is initiated in the leaves and the lesions develop on the stem by Myrothwium roridum. The disease is prevented and controlled with Copper oxychloride.

Earlier records indicate that old varieties were a susceptible to a virus causing the paling, crinkling and rolling of the leaves and the twisting and elongation of the petiole, with the crowding of leaves at the apex. The pathogen is transmitted by whitefly, Bemisia tabaci. Cotton seems to be collateral host.

The important pest of H. sabdariffa var. altissima is mealy bug, Macollicuus (=Phenaccus) hirsutius Green, causing bunchy top. Methyl demeton (0.05), Parathion (0.04), Monocrotophos O.04) and Fonitrothion (0.04) control the pest effectively.

The important diseases are:

- Root-and stem-rot caused by Phytophthora paracitica F. sriffae. The disease is controlled by drenching the soil and the basal portion of the crop with Copper oxyechloride and by ensuring good drainage.
- ln areas with high rainfall, leaf-rot caused by Phoma salwa-riffae is severe, but does not affect the fibre directly. The polyphagous hectic, Podagria apicifulva Bry. feeds on leaf-blades and an intense attack leads to damage in the case of HC and HS, more so in the former. HC 10 per cent prevents and controls the pest.

Released Varieties

HS varieties in use are 'RT 1', 'RT 2', and 'PT 26'. The new released improved type 'US 4288' has bristled green stem with a flash of released pigment at nodes, slightly proliferating to internodes at maturity; the leaves are deeply palmatised with 5 long narrow lobes; the petiole is pigmented on the upper surface only. The petals are sulphur yellow, with deep-purple throat. The variety flowers in late-November, irrespective of the date of sowing. It attains a height of about 400 cm. The duration of the crop is 180—210 days. The weight of 1,000 seeds is 18 g. The seeds are coffee brown. The yield potential of tile variety is 32 quintals per hectare. With average management, the yield is about 27 quintals. The variety is suitable for areas north of 22 °N.

'AMV 1' has bristled stem, with crim, on-purple pigment, both at the nodes and internode, the upper part of stem being more intensely pigmented. The leaf is deeply palmatisect (as in 'US 4288'); the petiole is red all through; the petals are sulphur yellow with a margin. The height attained is equal to that attained by 'HS 4288'. The variety is suitable for dry rainfed regions, specially those south of 22 °N.

HC varieties in disuse are 'MT 15', 'KT 129', 'Mt 150' and the NP series. The new improved type 'HC 5M' has, more or less, smooth green stem with an irregular light flush of red pigment, the pigmentation increasing with maturity. The leaf is entire, cordate and flowers are large, the sepals are green, with red spots and each with a prominent neetary (gland); the petals are lemon yellow, with deep crimson throat. Seeds are greyish black, more angular. A thousands weigh 32 g. The variety is quick-growing and its attainable height is between 400 and 500 cm in warm humid regions, and less in drier areas.

Promising Varieties

The non-bristled stem, equals 'HS 4288' in yield and the cultivators prefer it to the bristled types. 'HC 867' has excelled 'HS 583' in yield at places and possesses better resistance to the spiral borer, A. acutus.

Plant height, base diameter of the stem and the yields of US and HO have the same correlations as in the case of jute.

Quality and Grading

The quality of the mesta fibre is judged on the basis of almost same criteria as in the case of jute and grading is done on the same lines, although the I.S.I.sp. are now due. Natural cross-pollination being high the Varieties of the same species may be grown seed close at the same farm or near the farmers crop.

Ramie

Ramie (boehmeria nivea (L.) gaud) is a shrub belonging to family Urticaceae. It yields from its stem from the strongest vegetable fibre that makes excellent raw material for blending with terylene or rayon and itself can be spun and woven into excellent cloth that takes brilliant colours of dyes. The fibre goes into the making of sophisticated paper, including currency blanks. Though ramie requires special textile machinery, it can be adapted to the existing cotton and silk-waste machinery.

The technology for raising and running ramie plantations has been developed (Ramie in India: JARI). Since a viable plantation requires a large investment and at least 40 hectares, small holdings are thought to be unfit for commercial production. An alternative can be an agglomeration of co-operatives in a compact area.

Ramie requires a warm and humid climate for proper growth. It cannot tolerate water-logging and does not thrive at a low temprature. It requires well-distributed rainfall around 1,600 to 2,500 mm. North of Bengal, Assam, the foothills of Meghalaya and the Western Ghats are suitable for its cultivation.

Ramie is unisexual monoecious, and under certain conditions pseudodioecious, and cross pollinated. Its Rhizomes can be used in its propagation. A plantation is raised from rhizome cuttings (two quintals to cover 1 ha) placed 50 cm apart in rows 1 m apart; space variation is permissible. The planting time in north-eastern India is April to June. The new improved varieties suitable for northern India are 'R 1449', 'R 1452', 'R 1411'. Diseases and pests at the moment have not been reported.

At each point of planting, a clump of 8-12 canes (stems) develops. The first years growth being not uniform is cut back close to the ground in April. From the second year, depending on the growth, 4-5 cuttings at the harvest are possible at about 55-day intervals yielding about 12-16 quintals under gummed dry fibre per hectare per year. This yield is commercially attractive. Ramie, being a semi-perennial crop, is replanted in cycles of 7 to 8 years, applying heavy quantities of organic matter. Each cutting, except the last one, is to be followed by an application of ramie-west compost and fertilizers (NPK). Acid soils are to be amended by liming, whereas sandy and clay soils are unsuitable.

Ramie is not retted like Jute or mesta. Extraction is done directly from fresh canes with raspador decorticators indigenous or imported. Extracted fibre retains some gum and markatable as such. The removal of gum is done by consumers by using various techniques, including boiling with alkali using the tri-polyphosphate buffer.

Some Indian textile mills have found in ramie a good source of textile raw material. However, organized plantations are not many. This matter requires the attention of industrialists. There is a great potential for the export of ramie.

Sisal

Sisal (agave sisalana perrine), introduced from Mexico, is a semi-perennial plant and yields cream-white fibre from its leaves. The fibre is used for making ropes, matresses, brushes, doormats and fancy articles of various kinds. Moribund plantations of sisal are found in southern Bihar, Orissa, Andhra Pradesh, Karnataka, Maharashtra, Madhya Pradesh and West Bengal, raised in the early part of this century.

The technology for raising and maintaining a sisal plantation in India has been developed (sisal in India-JARI). As a result old plantations are being renovated and new ones are being planted on scientific lines to cater for commercial needs. The planting material can be obtained from the Sisal Research Station, Bamra, and from other plantations, particularly in north-western Orrisa. The plant popularly known as century plant, has a short thick stem (bole) which is almost concealed in early stages by the surrounding sessile leaves which are thick, long and broad. From slender rhizomes, buds grow and sprout above-ground to form baby-plants or suckers. After 8-10 years the plants flowers. The stout floral axis develop slowly and produces flowers which drop off and bulbils are small baby-plants develop on the seat of flowers. Both bulbils and suckers furnish the propogation material for plantations.

Bulbils and suckers are nurtured with copious nutrients in nutrients for a year in close spacings and later transplanted when 10 to 18 months old. Depending on growth the harvesting of the leaves starts from 2 to 3 years old plants and such annual harvests last for 7 to 8 years comprising a full cycle. The fibre is extracted with power-driven decorticators. Raspador decorticators are being manufactured in India. After a cycle is completed, the plantation is cleared, heavily manured and is preferably put for a year under a green manure crop and then replanted.

The yield of two tonnes per hectare per year may be considered good in the dry, infertile areas of north-western Orissa or Chhota Nagpur. With good management, the yield can be much improved, more so in better types of soil, with a rainfall between 650 and 1300 mm. Good drainage is essential. Gravely or highly acidic soils, poor in nutrients, are unfit for fibre production.

When exposd to rainfall above 1500 mm, sisal leaves become susceptible to a number of fungal pathogens, including colletotrichun, diplodia and alternaria spp. in the wilting of plants has been observed, but not frequently.

Sisal suffers most from imbalance in nutrition. It requires large quantities of calcium, magnesium and potassium, besides nitrogen and phosphorus. In acid soils, with pH below 6, this imbalance begins to be reflected in various kinds of symptons and the remedy lies in the application of soil correctives, organic manures and fertilizers. Phytophthora and Pythium spp. have been reported on exotic varieties introduced recently. This is a serious disease and planters must use only indigenous resistant types. With assured internal consumption, the present target is 8000 hectares. There is scope for many more plantations. The other two species found to have been used in India for plantations are Agave cantala Roxb. and Agave veracruz Mill. The fibre of these two species is often marketed as sisal fibre.

Sunnhemp

Sunnhemp (Crotaria junea L.) or Bombay hemp or Banaras hemp, is a crop grown for the bast fibre it yields. It is also a green-manuring crop, but a minor fodder crop of minor importance. The fibre is dull yellow somewhat coarse, strong and durable. Owing to the fibre having a high cellulose content, low lignin and negligible ash, the paper industry has identified it as the most suitable indigenous raw material for manufacturing tissue paper and paper for currency. In rural areas, it is also used for making ropes, twines and nets. It is also used for making canvas and screens (iat-patti). Sunnhemp fibre is not used for textile purposes, unlike jute and mesta.

The area under this fibre crop is 1,38,000 hectares and the annual production of fibre is 50,000 tonnes.

Climate

The crop is grown in almost all parts of India. But the states of Uttar Pradesh and Madhya Pradesh are most prominent in this respect. The crop grows best in tropical and subtropical climate. It is grownas kharif in the northern states. In the southern regions, Where the climate is, more or less equable and the winter is not pronounced, it is raised in rabi also. A minimum of 40 cm of rainfall distributed in not less than 50 rainy days during the growing season is a primary requisite.

Soil

Well-drained alluvial soils, having a sandy loam or loamy texture are suitable. For uncultivated fallows or freshly reclaimed soils it is an ideal first crop. Being a leguminous crop, the nodule formation depends on the calcium and phosphate status of the soil, hence, old alluvial soils with low pH are not suitable, unless acidity has been corrected by liming.

Cultivation

The sunnhemp crop sown in the kharif preceeds rabi crops, *e.g.* wheat and oilseeds. It is sown broadcast, but should better be sown in rows. In broadcast sowing, the seed-rate is about 25 kg per hectare; in lines (30 line-to-line spacing and 5 to 7 cm plant to plant spacing), the seed requirement is only 15 kg per ha.

Since it is a crop that fixes its own nitrogen fertilizers are added. Doses of P and K, ammounting to 20 kg per ha is beneficial. Trases of boron and molybdenum are required, but these elements are not generally added, since most soils already contain them. Calcium is required in large quantites. The traditional sunnhemp belt of Uttar Pradesh, *i.e.* Banaras, Pratapgarh, Sultanpur, Azamgarh, Deoria; etc. do not require liming. After the application of the fertilizer during the preparation of the land, sowing follows immediately. After sowing, the soil is raked and laddered in order to put the seed 2-3 cm below the surface.

Sunnhemp seeds, not older than one year, generally retain 90 per cent germinability. The germination of sunnhemp seeds is epigeal. The seedings grow very fast and in the process, they smother all weeds, except nut sedge. Weeding is not, therefore, necessary. Yet a common weed associated with sunnhemp is Ipomoea sp., whose seeds are disseminated along with the sunnhemp seed itself. The Ipomoea plants are twiners and grow along with the crop neck-to-neck, flower and fruit at the same time as sunnhemp. Unless manually weeded out, they hamper the growth of the crop. Seeds of these weeds can be separated from sunnhemp saads with some care.

Harvesting

The fibre crop gets ready for harvest in 120 to 150 days. Pod formation is considered the proper stage of harvesting. The crop flowers from September onwards, and this period sychronizes with the onset of short days in latitudes around 23 °N to 24 °N, where the main sunnhemp belt lies. The flowering time for most of the varieties remains almost the same, irrespective of the date of sowing. The minimum vegetative period needed before the initiation of flowering is 45 days in the case of most of the varieties (*e.g.*,'k 12 Yellow').

But there is an early-maturing strain ('t-6'), which is also day-neutral and flowers in 30 days, irrespective of the sowing time. This type is generally harvested at the dead-ripe pod stage in order to yield both fibre and seed, as is practised by the farmers of Midnapore (West Bengal) and the adjoining districts in Bihar. It is characteristic of this type that even after the maturity of the pods, the stalk remains green and the fibre is extractable. In this case the total cropping period is 80 days. The plants are harvested with sickles and the bundles are kept in the field for 2 to 3 days for the leaves to be shed.

Retting and Extraction

The process of retting is similar to that of jute. Unlike the case of jute, the source of water for retting in the sunnhemp-growing areas is not abundant. In most cases, retting tanks are shallow muddy ditches adjoining the fields where the run-off water of a shorter spell of the monsoon in collected. The retting time needed by sunnhemp is only 4 to 6 days.

The extraction of the fibre of sunnhemp is more cumbersome than that of the fibre of jute. The beat-and-jerk method is unsuitable in the case of sunnhemp as the tendency of the fibre to fibre to stick to the wood is more and the fibre gets entangled with broken twigs. Hence the extraction is done stalk by stalk from the lower end upwards. The fibre is washed in water and dried in the sun. Incomplete retting makes the fibre coarse and green, although its strength is not impaired. Inadequate washing leaves behind some gum which renders the fibre coarse. Paper-making concerns, which make a bid to produce a small chunk of their raw-material stock by themselves, resort to mechanical decorticated lump goes directly to the bleaching and digesting plant.

Varieties and Yield

Sunnhemp has a fibre content of 2-4 per cent on the basis of the weight of green twigs, or 8 to 14 per cent in terms of dry weight. This is a low figure contributing to low yields. The national average is such a low figure as 3 to 4 quintals of fibre per hectare.

The cultivated types acclimatized in the diverse ecogeographical belts exhibit variability to their vegetative growth, branching, leaf size and thickness, and more pronouncedly with respect to their sensitivity to photoperiod.

There were a number of improved varieties.

S.No.	State	Variety	Special features
1	Uttar Pradesh	'Kanpur-12' ('K-12')	High yielding, good-quality fibre resistant to wilt
2	Madhya Pradesh	'M-18'	Matures early; suitable for areas having light soil, with low rainfall
		'M-19'	same as 'M-18 plus moderate resistance to shoot-border
3	Bihar	'BE-1'	Good Yield and Quality
4	Madras	'Bellary'	High yielding
5	Maharashtra	'D-IX'	Wilt resistant
6	West Bengal	'ST 55'	Found yielding higher than K-12

Varieties under serial Nos. 1 to 5 were evolved before 1948 and No. 6 before 1958. Except the 'K 12' variety, much is not known about other varieties at present. 'K 12' has been the most widely-cultivated variety, which in course of tinle, an suffered from genetic mixing with poor-yielding local types owing to cross-pollination.

Later this variety has been further improved (1971) and released as 'K 12 yellow'. This is easily distinguishable from other types because of its yellow seeds in place of black ones of every other type. This variety has a high-yield potential up to 14 q/ha, and is moderately resistant to the wilt disease. The fibre is of good quality and lustre.

Seed and Seed Production

Sunnhemp seed occurs in shades of black in most cases and yellow in one case. The hundred-seed weight ranges from 3.4 to 6 g, depending on the variety. The statistics of seed requirement are lacking, since the massive quantity used up in green manuring is not accounted for. Sunnhemp seed is a Aonree of seed protein used for manufacturing adhesives.

The seed production of sunnhemp is yet to be organised by tradition, the farmer lows a few lines of sunnhemp on the borders of ragi and jowai or on the bunds of paddy fields for obtaining seed. The seed-crop is raised late in the kharif season, *i.e.* in early August, in the sunnhemp-growing districts of the northern states. But it is ideal to grow a seed-crop in the rabi season in a zone where winter is not severe, and the temperature seldom falls below 10 °C. Under optimum conditions, the seed shield will touch about 20 quintals per hectare.

They are:

- Xyllatipes Faber
- Xylocopa fenistroides Faber (the bumble-bees)
- Mogachile lanata fibre (111) belonging to the bee family.

Sunnhemp for Green Manuring

A sunnhemp crop Grown thick by using a heavy seed-rate (60 kg per hectare) with profuse branching and foliage is ready for incorporation into the soil within 2 to 3 months of sowing. At this time, the stalks will be tender with very little fibre formation, and hence, easy for quick disintegration in the soil. The green matter at harvesting weighs kbroun 20 tonnes per hectare. It contains about 80 kg of N per hectare, besides P and K. The crop is either ploughed in it stands or is cut close to ground and ploughed down and levelled with a wooden plank (sohaga). In the case of double-cropped paddy, the green-manuring crop is cut and carted to the field and is incorporated into the soil. Green-manuring with sunnhemp is practised invariably in every type of soil. It is particularly found suitable in reclaiming saline alkaline soils. The other green-manuring crop used is dhaincha (Sesbania sp.).

Diseases

Important diseases are wilt (Fusarium udum) mosaic (virus) spread by Bemisia tabaci; anthra-e-nose (Colletotrichum curvature); rust (Uromyres decoratus) and powdery mildew (Oidiuin sp.). Good management, crop rotation and seed-dressing can control these diseases.

Pests

The important pests of sunnhemp are:

- Top-shoot borer (triuntra Meyr) whose larvae burrow into the growing tips arresting their apical growth add inducing branching;
- The hairy caterpillar whose larvae feed on the foliage and the pods.

Spraying with Diazinon (0.04%) 0411 control the former, whereas 1% B.E.A. or Methy) Parathion (0.04%) can control the latter.

Flax

Flax or linwed (Linum usitatimsutn L.) fibre come from the stems and the oil from the seeds. The fibre obtained from the linmwda is inferior in respect of length and quality. There are some types exclusively for fibre. The flax fibre is valued due to its strength and durability which are superior to those of cotton. It is soft, shining and poeon high water absorbency. It has low elasticity and is stronger when wet than when dry. It is resistant to high temperatures, moisture and mudow. It is not readily reaettomordants and DYM, unlike cotton. B'abrim made of flarlaunder well. Max contains 80 to 90 per cent cellulose.

Flax in woven into fine fabrics, Much on lawns, cambric and drills, canvas and buckrarns. It is used in linen, stitching, making twines and nets for fishing, ropes, carpet- backing, wrapping cloth, and house furnishings. It is a good raw material for tissue paper, fire-fighting hose and water-bag. Although in the Gangetic flax types introduced by Italy, were not found the bre amp; on the contrary, they deteriorated into the linaced types. The attempt to produce a dual-pur, *i.e.* strain, have so far failed. We are bout SW tonnes of flax to meet the demands, chiefly from the Defence Organization. India's own efforts to cultivate flax have not been successful because of the climate.

Climate

Flax is a crop of the temperate zone. A cool humid climate with temperature ranging from 10 to 27 °C and a diurnal fluctuation of relative humidity between 30 and 95 per cent, cloudy weather during the growing period is preferred. In the hot and dry climate, flax types tend to branch and grow as linseed types. The ideal geographical regions envisaged for successful flax production in India are the valleys of the Himalayan ranges where the winter is similar to European summer; this possibility has been proved by some trials.

Soils

Heavy clay and light soils are unsuitable. Rich loams or clay loams are considered best. The 'char' lands of the Ganges and the Brahmaputra which receive silt from the annual floods are suitable in winter.

Cultivation

For the Gangetic plains, the sowing time is early November. For the high range, it will be early October. The seed is drilled in rows 20 cm apart. After germination, the seedlings are thinned to keep a plant-to-plant distance of 1 to 2 cm. The wed-rate is 40 kg, the expected germination heirs 90 per cent. One or two irrigations will be needed. The crop matures in 100 days, one month after the initiation of the first flower. Harvesting is done by cutting the stems close to the ground or by pulling them out.

Retting

Bundles of stalks (straw) steeped in retting water just enough to cover. It take about 3 days for the retting to be completed under the normal temperature prevailing at the harvest time. Over-retting weakens the fibre; under-retting makes extraction difficult. Betted straw is then dried.

Extraction

The extraction of fibre requires a simple machine. The machine is indigenously designed on the, principle of passing the straw through fluted rollers to break the woody core and separate the fibre. The separated fibre strands are then brushed and rolled into bundles. This machine can also be used to salvage fibre from the linseed straw that go or at best can be used as fodder or can be burnt as a fuel. The fibre obtained way be inferior in quality, but it can be used for paper-making.

Varieties and Yield

Flax motivation of the put depended on varieties straightaway import from abroad. The recurring imports of wed in large quantities (owing to the deterioration of the crop) was not economical.

Cotton

Cotton is one of the most important commercial crops playing a key role in economic, political and social affairs of the world. Some new selections ('F 895' and 'F 896') have been made and they have become acclimatized.

Chiefly as a fibre crop cotton is cultivated in about 60 countries of the world but ten countries, *viz.* the USSR, the USA, China, India, Brazil, Pakistan, Turkey, Egypt, Mexico and Sudan account for nearly 85 per cent of the total production.

The organized sector of the Indian textile industry constitutes the largest single industrial segment in the country in terms of the annual value of output and labour employed, both direct and indirect. There are 688 cotton textile mills in India, as enumerated on 1 January 1973, having an installed capacity of 18.4 million spindles and 207 thousand looms.

The industry provides direct employment for nearly 900,000 workers and indirect employment for several millions. The decentralized sector comprising the power-loom, hand-looms, charkha, etc. is reported to provide -employment for over 2.5 million. With nearly 8 million hectares under the cotton crop, India ranks first in the world in respect of area and fourth in total production which -reached the level of 63.3 lakhs of bales during 1971-72. The other major cotton-growing countries of the world are the United States of America, the USSR and China.

Special Composition

India grows on a commercial scale varieties falling under all the four cultivated species of Gossypium, *viz.* G. hirauguin, G. barbadense, G. arboieum and G. herbacin. The predominant species cultivated is G. hirsulunt, which covers about 50 per cent of the area of 8 million hectares, followed by G. Ambwmm with 29 per cent and G. Aeum with 21 per cent. The area under G. barbanse is negligible and covers only a few thousand hectares. Cotton research and development efforts are directed towards progressively increasing the area under G. hirmaum and replacing, as far as possible, the area under the old world species. The cotton crop in the United States is now approximately 99 per cent G. hirsutum and one per cent G. barbanse In the USIT, G. hirmauits is grown over 92 per cent of the area and G. barbadenje over the rest of the area.

Area under Irrigation

Cotton in India is largely cultivated under rain-fed or barani conditions. It is mainly raised during the tropical monsoon season, although in the southern parts of the country, it in cultivated during the late monsoon season in winter. In nearly 75 per cent of the area, the cotton crop is entirely dependent on rainfall, whereas supplementary irrigation facilities exist in about 25 per cent of the area. There has been a progressive increase in the area under irrigated cotton in India during the past 25 years.

Year		**Area under cotton (in million hectares)**	**Percentage area irrigated**
	Total area	**Irrigated area**	
1947-48	4.26	0.29	7
1951-52	6.56	0.58	9
1956-57	8.05	0.93	12
1961-62	7.72	1.08	14
1966-67	7.83	1.23	16
1971-72	7.78	1.95	25

Climate and Soil

Cotton is tropical subtropical crop. For the successful germination of its seeds a minimum temperature of 15 degree is required. The optimum temperature range for vegetative growth is 21-27 °C. It can tolerate temperatures as high as 43 °C, but does not do well if the temperature falls below 21°C. During the period of fruiting warm days and cool nights with large diurnal variations are conductive to good boll and fibre development.

Crop Season and Rotation

The sowing season of cotton varies considerably from tract to tract and is early sowings done in the Dharwar- Gadag area in the beginning of September have been found to give better yields. In addition, summer sowings in Tamil Nadu is done during February-March.

The sowings of cotton in the rice fallows of Andhra Pradesh and Tamil Nadu extend from the second half of December to the middle of January. The modest rotations for cotton in the irrigated areas of the Punjab, Haryana, northern Rajasthan and Uttar Pradesh are cotton-wheat, cotton-jewar and cotton-wheat-toria. The growing of berseem and Guar has been found to have a beneficial effect on the succeeding cotton crop.

In the central and western India, cotton-jwar, cotton-bajra, cotton-wheat, cotton-gram and cotton-sesamum are the usual rotations followed. The raising of the cotton crop after a legume in rotation and preceded by a grain crop has been found to give a better yield. Cotton following groundnut has been found to give 15-20 per cent higher yield than that following wheat or jowar. Another common practice in cotton cultivation is mixed cropping with maize, jowar, ameboid, sesamum, pulses or vegetables as practised in many parts of central India. Intercropping with ragi, other millets or groundnut is also quite common in parts of Tamil Nadu, Karnataka and Andhra Pradesh. Intercropping and mixed cropping under rain-fed conditions serve both as an insurance against crop failures and as a preventive against soil erosion.

Preparation

For the irrigated cotton crop, following wheat in the northern states of India, the preparation of land is usually hurried through, since the interval between the harvesting of the grain crop in March- April and the sowing of the irrigated cotton in April- May is very short. The soil is first given a heavy irrigation and one or two ploughings. There after, a light irrigation is given followed by one or two ploughings and then the soil is planked with a wooden plank (sohwa) before cotton is sown. The black-soil areas of central and southern. India are prepared for sowing the rain cotton by harrowing- field 3 or 4 times with the blade harrow. The land is given a deep ploughing only once in 4-5 year to remove weeds.

In the red and lateritic soils of southern India, which are poor in retaining soil moistures, the field is usually given two or three light ploughings for raising cotton as a rabi crop. The irrigated crop in these soils is given one watering, ploughed once and the land is well for sowing the irrigated winter crop.

Seed-Rate and Spacing

For obtaining maximum yields, proper seed-rate and spacing are essential. Depending on the variety, the level of soil fertility and the cultivation practices, seed-rates and spacing and have bets recommended for different improved varieties.

A seed-rate of 15 to 25 kg per hectare and a spacing of 75 to 90 cm between the rows are generally recommended for cotton grown under irrigated conditions. For the rain-fed cotton, a seed-rate of 12 to 16 kg per hectare and a spacing of 45 to 60 cm between the rows are adopted. For the rain-fed American cotton, the seed-rate is 12 to 16 kg and the spacing is 60 to 75 cm between the rows, broadcasting the seed or sowing it in lines either with a mechanical-drill or by hand, are the two common methods. Line-sowing with wood-drills is recommended in order to ensure uniform germination, better stand and easy inter-cultivation.

In the case of the ridge-soak irrigated crop, hand-dibbling of the seed at the spacing is commonly followed. The method is also followed in dryland areas of parts of Madhya Pradesh and Andhra Pradesh, adopting the square-pocket system of cultivation.

Manuring

The Cotton crop in most of the areas in India is usually not adequately manured. The soils in which cotton is grown are generally sufficient in phosphate and potash, but since they are deficient in nitrogen and matter, it is necessary to apply them for better yields. Any leguminous crop green-manure crop or as a crop in rotation preceding cotton, increased the yield of cotton.

Farmyard manure is rarely applied to the cotton crop in the states of Punjab, Haryana, Rajasthan and Gujarat. Of late, sonic of the recreative farmer, states apply 6 to 12 tonnes farmyard manure per hectare to the rain-fed crop and 15 to 28 tonnes of it to the irrigated crop.

In Maharashtra, the cotton crop is manured with farmyard manure once in 3 or 4 years at the rate of 12 to 15 tonnes per hectare. In Tamil Nadu, the irrigated American cotton crop is usually manured heavily and supplemented with fertilizer.

The need for applying chemical fertilizers for increasing cotton yield has been amply proved. The usual dose recommended is 40 kg of nitrogen per hectare in the case of the irrigated crop and 20 kg in the case of the rain-fed crop. In northern India, a higher dose of 60 to 100 kg N/ha has been found to he economical. Good response to high doses of 80 to 100 kg

N has obtained in parts of Maharashtra, in the fallow an of Andhra Pradesh and in the case of winter-sown coton in Tamil Nadu.

The application of nitrogen by band placement has been foun to give better yield broadcasting. For the irrigated cotton crop, it has been found to be economical to apply half of the basal dressing at the tigne of sowing and the other half as top-dressing during thinning or just before flowering.

In the case of the irrigated and row-gown cotton crop, inter-cultivation is done fairly regularly, with either a blade harrow or with a three tined hoe or with a dui plough. In the case of the crop that is grown broad, one or two hand-booings are given to remove weeds. Inter-culitivation not only checks the growth of weeds, but also lead to better soil aeration and soil-moisture conservation.

The thinning of the cotton crop is a special feature of the irrigated crop sown on ridges in Peninsular and in parts of southern Gujarat. Manuring is desirable for maintaining the optimum population of plants to obtain a high yield. During this operation, the vigorous things are retained and weak and oil-type ones are removed.

In the case of irrigated cotton, the option moisture supply during the critical stages of growth and development is the most important aspect of cultivation. A proper irrigation schedule in the case of this crop in relation to soil type, weather conditions, cultivation practices and the variety is the decisive factor in getting the high yield.

The cotton is grown in variety of soils it require to amenable to good drainage as it does not tolerate water logging. It is grown mainly as a dry crop in the black cotton and medium black soils as an irrigated crop in the alluvial soils.

In Tamil Nadu, the major portion of irrigated and rain-fed crop is planted in August-September.

Year		**Area Under Cotton (in Million Hectares)**	**Percentage Area Irrigated**
	Total area(in Million Hectare)	**Production in Lakh**	**Lint Yield**
1947-48	4.26	0.29	90
1951-52	6.56	0.58	85
1956-57	8.05	0.93	105
1961-62	7.72	1.08	114
1966-67	7.83	1.23	157
1971-72	7.78	65.3	157

Well-drilled holls alone are picked, as otherwise the quality will deteriorate. It is better to remove the dry stalks an won as the picking is completed, as they usually harbour insect pests if they are left in the field. The yields vary widely from tract to tract and from to redounding upon the climatic condition. The unit area of cotton in India during 1947-48 was only 90 kg per hectare and it had been estimated at 151 kg/ha during 1971-72.

An-India Co-ordinate Cotton-Improvement Project

Main Improved Varieties

The All-India Co-ordinate Cotton-Improvement Project was sponsored by the Indian Council of Agricultural Research in 1967 and as a result of nitrification of research 16 new varieties of cotton were released during 1968-72 for cultivation. They are, 'Kriobna' for the rice fallow tract of Andhra Pradesh, 'MCU-5' -and 'Sujata' for the irrigated Cotton tracts of Tamil Nadu, 'Khandwa-l' and 'Khandwa-2' for the rain-fed of Madhya Pradesh, 'G-27' for the desi cotton tracts of Punjab and Haryana, 'J. 205' for the central sistriete of Punjab, 'Barathi' and 'JK.S' for the rain-fed Karunganni tract Tamil Nadu. 'Mabalaxwi' for the rain-fed northern tract of Andhra Pradesh, 'H Ibr' 'ld-l' cotton in Gujarat and other state, 'Sujay' is for the southern Gujarat tract, 'IMCU-7' for the medium-duration rice-fallowing of Tamil Nadu, 'M-89' for the Briganganagar tract of Rajasthan, 'Varalaxmi' hybrid for the Turigabbadbra Project area in Karnataka and 'GS-23' for the low-rainfall areas of Karnataka. Out of the above mentioned varieties the release of extra long-staple variety 'MCU-5' and 'Hybrid-4' and 'Varalaxmi' hybrid cotton constitute a sufficient mile- stone in cotton research. Variety 'Sujata' is the first high-spinning cotton released in India. The spread of 'Vyplbrid-4' cotton had a great impact on production and the quality of the Indian cotton crop has been significantly stepped by variety 'MGU-S'. Besides the above varieties have already been released, a number of promising varieties have been built up to meet the requirements of different zones in the various quality groups. Based on further evaluation they are expected to be released during the 1970s.

Pests and Diseases

The cotton crop is subject to damage by a number of pests right from the time of germination till the final picking. Very intensive work has been carried out on the cotton pests to evolve measures to control them and obtain economic yields. The pests of major significance in most of the cotton-growing tracts include aphids, jamids, thrips, bollworms, and mites, which affect the yield and quality of cotton. Ephylactic control schedules have been developed for checking the damage from pests and they have formed the basis for the crop-protection programmes in cotton cultivation.

Diseases of cotton cause reduction in yields by affecting germination, by killing the seedlings, by lowering the productivity of plants and also by affecting the quality of lint. Root-rot and wilt, caused by several fungi, bacterial blight and anthraenose are the major diseases of cotton in India. Control measures, including seed treatment, soil-drelaches, spraying the foliage with fungicides and the use of resistant varieties have been developed for the major diseases.

Processing and Technological Evaluation

Ginning

The ginning of seed-cotton occupies an important place in the processing of material from the field to the factory. The act of separating the fibres from the seed cotton is known as ginning. This can be done either with roller-gins or with saw-gins. In India, the gins used are mostly roller gins.

Cotton Quality Evaluation

First spun into yarn before being used ultimately to manufacture fabrics. Hence, the price paid for a sample of cotton is related to the quality of yarn which can be spun from it. In India, this spinning performance is usually expressed as the 'Highest Standard Count' (HSC) to which it can be spun. The fineness of yarn is usually expressed in terms of counts. A count is the number of hanks of 840 yards each, which weigh one pound; for instance 20 counts of yarn would indicate that one pound of a particular yarn would contain 20 hanks of 840 yards each. Based on the end uses of the yarn, certain strength or standards have been prescribed for different counts by the Cotton Technological Research Laboratory of the Indian Council of Agricultural Research.

The Highest Standard Count for cotton is that finest count of yarn which can he spun to satisfy the yarn-strength standards. The important fibre characteristics which determine the quality are the fibre-length parameters, fineness, maturity and strength. Other properties, such as convolutions, rigidity and inter-fibre friction, play only a minor role.

In addition to the above, two other characteristics, viz:

1. The colour and appearance of cotton,
2. The trash content are also taken into account for the fixation of price in the market.

Fibre Length Parameters

Fibre length is one of the most important characteristics and the market price is determined to a large extent by the staple length. Other factors being equal, longer linted cottons give better spinning performance than shorter linted cotton.

Staple-Length Classification

Based on staple length, the cottons produced in various countries of the world are classified into three main groups, *viz.* short staple (9.5 mm-25.4 mm), medium staple (13 mm-40 mm) and long staple (24 mm-63 mm) which are somewhat overlapping each other.

The staple-length classification of cottons adopted in India is not identical with that adopted in the USA or other foreign countries, but is based on indigenous production. The groups now recognized in India are superior long staple (27 mm), long staple (24.5 mm-26 mm or 31/32 inch-33/32 inch), superior medium staple (22 mm-24 mm or 28/32 inch-30/32 inch), medium staple (20 mm to 21.5 mm or 25/32 inch-27/32 inch), and short staple (19 mm or 24/32 inch and below.

The Cotton Making in India

The marketing of crops close of harvesting of kapas end after the lint is produced by the mills. Between the set points, it passes through several stages, *viz.* sale as kapas in the primary and secondary marketing, ginning and processing, storage, transport to terminal markets and the sale of lint to the consuming mills.

In primary markets, kapas is sold by the grower to the village merchant or itinerant merchant without the intervention of any intermediaries. However, in recent years direct sales in the village have dwindled considerably. The bulk of the growers, nosy dispose of kapas in the secondary markets, *i.e.* important trade centres. Here, the produce is bought by stockists, ready dealers, brokers or commission agents and by the mills themselves which own ginning factories. In the terminal markets, cotton lining old to the textile mills, exporters and traders dealing with the consuming mills or engaged in inter-state trade. Bombay, Coimbatore, Ahmedabad and Kanpur are some of the important terminal cotton markets of which Bombay is the largest.

Apart from the above system of marketing, farmers co-operative cotton sales societies are functioning in many states. The first co-operative cotton sales society was opened at Gadag in Mysore in 1917. Later, such societies have been established in other parts of the country and at present co-operatives not only undertake the marketing of cotton but also undertake its ginning and pressing.

The functions of these co-operatives are to provide the grower-members with credit, supply agricultural inputs to them, and give it to the mills.

Monopoly Purchase in Maharashtra

The Government enacted a scheme of monopoly state under the Maharashtra Raw Cotton (Procurement, Processing and Marketing) Act, 1971.

The Act has three objectives, viz.:

1. To ensure that the farmers get a fair share in the market price,
2. To supply unadulterated cotton to consumers at reasonable prices,
3. To guarantee the purity of cotton and honest trade practices at the processing centres.

Under the scheme, at the commencement of every cotton season, the state Government shall, in consultation with the Central Government, fix guaranteed prices for different varieties or grades of kapas and purchase them from the growers through collection centres. Trade in kapas is prohibited. The farmer gets 80 per cent of the guaranteed price as 'advance' on the delivery of kapas and the balance on the disposal of lint and seed by the Government.

The Cotton Corporation of India has also entered the marketing field of internally produced cotton from 1971-72 onwards.

The operations of this Corporation are, however, mainly confined to limited purchases to solve the problems of marketing.

Seed Multiplication

Specific schemes for the multiplication and distribution of the pure seed of improved strains had been financed by the former Indian central cotton committee in collaboration with the state government since 1929. Further sponsored schemes for the production of nucleus and foundation seed of all the improved varieties of cotton released under the research programme.

The subsequent stages of the multiplication of improved seeds are handled by the departments of agriculture through co-operative societies and the National Seeds Corporation, with the assistance of seed-growers.

The Central State farms are also involved in the in the production of pure seed cotton. Seed multiplicatton 1950 onwards, the Government of India has sponsored Cotton Extension Schemes in different states under these schemes, the distribution of good seed was one of the items. The work hitherto carried out under the above two schemes was later integrated under the Comprehensive Cotton development Scheme.

The Central Farms and the state seed farms are also involved in the production of the pure seed of cotton varieties.

Cotton Development Programmes

The Government of India has been implementing a number of schemes in different cotton-growing tracts for increasing production. The government is involved mainly and an Intensive Cotton District Programme has been implemented in selected districts, based on the area approach to crop development.

The Directorate of Cotton Development under the Department of Agriculture, Government of India, supervises the cotton-production programmes in the states which implement the cotton development schemes. The Indian Cotton Mills Federation's, Cotton Development and Research Association is financing a limited programme of research and extension activities.

The yield data from the farmers fields in the Government Sponsored Package Programme areas as well as from the Development Project areas supervised by the Indian Cotton Mills Federation have indisputably proved that much higher yield of cotton than what is obtained at present is quite possible through the adoption of improved practices.

Impact on Cotton Production

Consequent on the release and spread of the better yielding varieties of cotton in the various quality groups and the adoption of improved agronomic practices and crop-protection schedules, there has been a progressive improvement in cotton production in the country both quantitatively and qualitatively during the past 25 years, as may be seen from the figures set out in Table.

Year	Total	Long Staple	Medium Staple	Short Staple
1947-48	21.6	6.09 (28.2%)	11.93 (55.2%)	3.55 (16.4%)
1951-52	31.0	9.80 (31.6%)	12.08 (39.0%)	9.0 (29.3%)
1956-57	31.0	9.80 (31.6%)	12.08 (39.0%)	9.0 (29.3%)
1956-57	46.7	7.68 (16.4%)	19.83 (42.5%)	19.14 (41.0%)
1961-62	44.6	8.07 (18.1%)	17.47 (39.2%)	19.03 (42.2%)
1966-67	49.7	7.48 (15.1%)	30.77 (61.9%)	11.48 (23.1%)
19671-72	65.3	16.01 (24.5%)	39.65 (60.8%)	9.61 (14.7%)

Staple Length Classification

	Up to 1961-62	After 1961-62
Long	14-mm and above	24 mm and above
Medium	18 mm to 21 mm	20 to 24 mm
Short	17 mm and below.	19 mm and below.

Owing to the interaction of three major factors, *viz.* the availability and spread of new technology, favourable weather conditions and adequate financial resources with the farmers, because of the higher prices for cotton during the previous season, the country achieved a record production of 65.3 lakh bales of cotton during 1971-72 from an area of 7.78 million hectares. The highest production previously achieved during the most favourable weather of 1964-65 was 56.6 lakh bales from an area of 8.27 million hectares, thus bringing into focus the role of new varieties and improved technology in increasing production during 1971-72.

Tobacco

The genus Nicotiana has more than 60 species, of which two are commercially cultivated for the production of tobacco. They are N. Tabacum and N. rustica. N. is widely cultivated in most countries of the world, and as far north as Sweden, and as far south as New Zealand. The cultivation of N. rustica is restricted to India, Russia and some other Asiatic countries.

Though the tobacco plant is tropical in origin, its production at present is concentrated mostly outside the tropics, except in India. In tobacco production, India ranks third after the U.S.A. and China, with an annual production of 380 million kg of leaf from an area of about 445 thousand hectares. Almost all states in India grow tobacco, but the important ones are Andhra Pradesh, Gujarat, Tamil Nadu, Karnataka, Bihar, West Bengal and Uttar Pradesh in that order, and account for about 90 per cent of the annual production.

The species N. tabacum is grown in almost all the states, whereas the cultivation of N. rustice is confined to the northern and north-eastern states, where the temperatures are considerably lower during the growing season. India ranks fourth in the world in tobacco' exports, after the USA, Turkey and Greece, Pride exports about 80 million kg of tobacco to about 50 countries and earns about 635 million rupees of foreign exchange annually. About 85 per cent of the export consists of flue-cured Virginia tobacco, and India is the world's second largest exporter of this variety after the U.S.A.

For the improvement and development of different types of tobacco grown in the country, the Indian Central Tobacco Committee was established in 1945. In 1947, this Committee established the Central Tobacco Research Institute, Rajahmundry (Andhra Pradesh) for fundamental reference on all tobaccos and applied work on tobaccos grown locally.

Subsequently, the committee had taken over the research station at Guatur (Andhra Pradesh) from the then Industrial Agricultural Research Institute, New Delhi, and started the Regional Research Station at Yedasandur (Tamil Nadu), Hunsur (Karnataka), Pusa (Bihar) and Dinhata (West Bengal) for the improvement of the types of tobacco grown in these states.

Besides the above stations, a number of schemes run by the state governments for the improvement of the crop were financed by the Committee. With the abolition of the Commodity Committees in 1965, the Central Institute and the regional research stations attached to it came under the control of the Indian Council of Agricultural Research and the development activities were taken over by the Ministry of Agriculture, Government of India.

Soil and Climate

India produces a wide range of commercial types of tobacco and each state specialises in the production of specific types. However, all these types are mostly used for internal consumption except the flue-cured Virginia tobacco which is exported. The flue cured Virginia tobacco is normally grown in other countries during the rainy season in light soil with a sandy surface, a friable sandy clay subsoil, low in organic matter content and acidic in nature.

In India, on the other hand, it is mostly grown in heavy black-soils of Guntur, Prakasam, Krishna, Bust and West Godavari, Khammam and Nellore districts of Andhra Pradesh in winter.

Cultivation

Usually, 6 to 10 ploughings are given by way of preparatory cultivation. Experiments have shown that deep ploughing of the fields once in two years during summer, followed by 2 to 3 ploughings and 1 to 2 harrowings are beneficial for most of the tobaccos, particularly for the flue-cured virginia tobacco grown in Andhra Pradesh. Digging with a spade, followed by ploughing with a mould-board plough and a country plough and then a harroeing is recommended for Bihar.

Farmyard manure is usually applied and the dose varies from 10 to 125 cartloads per hectare for different types of tobacco. In addition, oil-cakes or ammonnium sulphate was found beneficial for some tobaccos.

Based on the research findings of the Central Tobacco Research Institute, Rajahmumdry, and its regional stations, the following mannual schedule is recommended:

1. Seven-and-a-half tonnes of farmyard manure, six weeks before planting and 22 kg of N + 50 kg of P_2O_5 + 50 kg of K_2O per hectare, a week before planting for the flue-cured Virginia tobacco on black soils of Andhra Pradesh.
2. Ten to twelve tonnes of farmyard manure, six weeks before planting and 30 to 40 kg of N + 80 to 100 kg of P_2O_5 + 80 kg of K_2O + 12 to 15 kg of MgO, of which the entire quantity of Mg and P_2O_5 and half the quantity of N and K_2O are applied a week before planting and the remaining quantities of N and K_2O are applied 3 to 4 weeks after planting as top-dressing just before irrigation for the flue-cured Virginia tobacco on light soils of Andhra Pradesh.
3. Fifteen tonnes of farmyard manure or 300 kg of groundnut-cake 2 to 3 months before planting and 44 kg of N per hectare a week before planting for natu tobacco in Andhra Pradesh.
4. Forty-five to 90 kg of N per hectare, half as oil- cakes and, half as ammonium sulphate for the bidi tobacco in Karnataka.
5. One hundred and eighty kg of Wilha, half as castor-cake or groundnut cake and half as ammonium sulphate, 2 weeks before planting for the bidi tobacco in Gujarat.
6. Twenty five tonnes of farmyard manure and 75 kg of N/ha (213 as ammonium sulphate and 113 as groundnut cake, half before planting and the remaining half 6-7 weeks after planting for cigar and cheroot tobaccos of Tamil Nadu.
7. Sixty tonnes of farmyard manure and 112 kg per ha, half as mustard-cake and half as ammonium sulphate for the chewing tobacco in Bihar.
8. Thirty tonnes of farmyard manure and 150 kg of N + 150 kg of P_2O_5 per hectare for hookah and chewing tobaccos; 17 tonnes of farmyard manure and 125 kg of N + 112 kg of P_2O_5 + 224 kg of KaO per hectare for the wrapper tobacco in West Bengal.

Spacing

The tobacco seedlings are always planted in rows. The distance between the rows and between the plants within the two varies with the type of tobacco. Based on the research findings, a spacing of 80 cm × 80 cm for natu and the flue-cured Virginia tobacco in black soils and 100 cm × 60 cm for the flue-cured Virginia tobacco in light soils of Andhra Pradesh; 75 cm × 50 cm for cigar and cheroot tobaccos and 75 cm × 75 cm for the chewing tobacco in Tanil Nadu; 100 cm × 70 cm and 105 cm × 45 cm for the bidi tobacco in Gujarat and Karnataka, respectively; 90 cm x 60 cm for the chewing tobacco in Bihar, 60 cm × 45 cm for N. rustica and 90 cm × 90 cm for N. tobacum in West Bengal, are recommended.

Interculturing

When after the plants get established, *i.e.* about 20 days after planting, the tobacco fields are given frequent inter-culturing and weeding to conserve; the soil moisture and cheek the growth of weeds. Tined harrows, blade harrows and the country plough are normally used for interculturing. Mulching the crop with paddy straw at 3,600 kg per hectare after the first interculture was found beneficial in the case of the flue-cured Virginia tobacco on black soils of Andhra Pradesh in increasing the yields and improving the quality by understanding soil moisture.

Irrigation

In Andhra Pradesh, flue-cured Virginia tobacco on black soils is not normally irrigated, but the crop on light soils is given up to six irrigations. The irrigation water should not contain more than 50 ppm of chlorides, as otherwise the leaves get burnt and other qualities suffer. In black soils also, in adverse conditions, one irrigation on 40-day-old plants is recommended. The cigar, cheroot and chewing tobaccos in Tamil Nadu and the chewing tobacco in Bihar are normally irrigated. The bidi tobacco in Karnataka, Maharashtra and Gujarat and the hookah and chewing tobaccos in West Bengal are not normally irrigated, but in years of deficient rainfall, one or two irrigations are desirable to maintain the yield.

Topping and Suckering

The removal of the flower head alone or along with some of the top leaves of the plant is known as topping. After topping, the axillary buds grow and their removal is known as suckering. Most of the types of tobacco grown in the country are topped and suckered for improving the size, body and quality of the leaves, except wrapper tobacco where the texture is one of the important criteria. The flue-cured Virginia tobacco grown in the black soils of Andhra Pradesh was also not topped till recently, but of late, topping is recommended to improve the body of the leaf. Even now, if the growth of the crop is rank and the colour of the leaf is dark green, topping is avoided. All other types of tobacco are topped. The level of topping varies from type to type and ranges from high topping in the case of the flue-cured Virginia tobacco, in which case only the flowerhead is removed, leaving all the 20-24 leaves intact. To the very low topping in the case of chewing tobacco of Tamil Nadu where only 7 to 8 leaves are retained on the plant. Based on the research findings, topping at 14-16 leaves for the cigar and cheroot tobaccos in Tamil Nadu, at 13-15 leaves for bidi, tobacco in Karnataka and Gujarat and the chewing tobacco in Tamil Nadu and at 8-9 leaves for N. tabacum and at 6-8 leaves for N. rustica in West Bengal, is recommended.

To get the full benefit of topping, the suckers are to be removed periodically. This is a laborious and time consuming job. Hence, a number of experiments for the suppression of suckers by the application of oils, chemicals, hormones, etc. were conducted and it was reported that the application of coconut-oil to the top six leaf axils soon after topping suppressed the suckers in the flue-cured Virginia tobacco. Two per cent naphthalene acetic acid triethenolemin applied to the topped portion of the steni suppressed the majority of the suckers in cigar and cheroot tobaccos of Tamil Nadu. In Bihar, though vegetable oils and hormones suppressed the suckers by 34 and 47 per cent respectively, in the case of chewing tobacco, they did not prove to be of much practical importance to the cultivators. However, piercing the stem after topping improved the yield and quality and is widely practised. Naphthalene acetic acid and Esso oil did not produce significant effect in controlling suckers in the case of the bidi tobacco at Anand.

Harvesting, Curing and Fermentation

The signs of maturity and the methods of harvesting differ with the type of tobacco. The flue-cured Virginia tobacco is harvested during December-March in Andhra Pradesh and during July-September in Karnataka. The leaves are considered ready for harvesting when the normal green colour changes to yellowish green or to light yellow. Harvesting starts from the bottom and each time 2 or 3 leaves are harvested. In another week, the next 2 or 3 leaves mature when they are harvested. Thus the 20-24 curable leaves, that become available per plant in the case of the flue-cured Virginia tobacco, are harvested in 6 to 8 primings at weekly intervals. Soon after harvesting, the leaves are strung on bamboo sticks at the rate of about 100 leaves per stick and loaded in the barn for curing.

The bidi tobacco is harvested in January-February when the majority of the top leaves develop red rusty spots known as spangles. The cigar and cheroot tobaccos are harvested 90 to 100 days after plant when the leaves pucker and become brittle and yellowish-green. The chewing tobacco is harvested 110 to 120 days after plant when the leaves develop Pronounced puckering. The hookah tobacco (rustica) is harvested in May or June, and the tabacum harvested when broad flecks appear on the leaves. The whole plants are harvested in the case of the bidi, cigar and cheroot, chewing and hookah tobaccos except in West Bengal where harvesting is done by priming.

Tobacco leaves are cured after harvesting in order to impart the required colour, texture and aroma to the final product. Different methods of curing are adopted for different types of tobacco, depending on its quality requirements and the use to which it is put to. Flue-curing, air-curing (or shade-curing), sun-curing (or rack-curing or ground-curing), smoke-curing and pitcuring are the different methods of curing in vogue in the country. The Virginia tobacco is cured in special chambers, known as barns with artificial heat passing through metal pipes, called flues. Hence this tobacco is known as the flue-cured Virginia tobacco and is exclusively used for manufacturing cigarettes. Curing takes 100-120 hours, after which the leaves are unloaded from the barn, bulked for a few days and then graded, depending on the colour. Since the quality of the leaf depends, not only on colour but also on its position on the plant, the plant-position grading is now recommended.

The wrapper tobacco of West Bengal and the Lanka tobacco of Andhra Pradesh are air-cured. The wrapper tobacco is cured under atmospheric temperature and at 70-80 per cent relative humidity in barns with sides and the roof closed. The Lanka tobacco is cured by hanging the strung leaf in thatched sheds for 2 to 21 months after which the curing is completed in pits.

Most of the tobaccos grown in the country are suncured, as in this method the construction of costly structures can be avoided and also because the sun is quite-bright and hot during this part of the year. The cigar and chewing tobaccos of Tamil Nadu are cured in the sun for 15 to 20 days by stringing the whole plants on bamboo poles. The natu tobacco of Andhra Pradesh is cured by hanging the Strung leaves on sea golds in the open for 1 to 11 months. In the case of the bidi tobacco, the plants are cut and left upside-down on the same spot for 3 to 7 days. After the leaves are dry, the lamina is stripped off. The chewing tobacco in Bihar is cured in the sun, the cut plants in the field for 4 to 7 days. In Uttar Pradesh, the plants are cut and exposed to sun for 1 or 2 days, after which the leaves are separate and spread on the ground or made into heaps for a further drying. In West Bengal, the leaves are primed and kept on the ground upside-down for a few hours for drying, after which they are tied into bunches of 4 to 5 leaves and cured in sheds for 4 to 5 weeks.

Some of the chewing tobaccos grown in Tamil Nadu are smoke-cured in cylindrical buts with open tops. The cured leaves, except that of the bidi tobacco are then bulked into heaps of convenient size for fermentation which improves the quality and aroma of the leaf. The bulks are broken up and remade occasionally to avoid excess moisture and decay. The leaf is then graded and packed, as required by the trade.

Yield

The average yield of tobacco leaf per hectare is about 750 kg and 950 kg for the flue-cured Virginia and the natu tobacco respectively in Andhra Pradesh; 1,000 kg, 450 kg and 350 kg for the bidi tobacco in Gujarat, Maharashtra and Karnataka states respectively; 1,250 kg and 1,600 kg for the cigar and the root and chewing tobaccos, respectively in Tamil Nadu, and 950 kg, 850 kg and 800 kg for the hookah and chewing tobaccos respectively in Bihar, West Bengal and Uttar Pradesh.

Varieties

There are local or commercial varieties grown in the various tobacco-producing areas under both N. tabacum and N. rustica. It is with the help of 'the Central Tobacco Research Institute, Rajahmundry, and its regional stations and other organizations connected with the crop.

6

Agricultural Marketing in Developing Countries

NGO AND CBO INVOLVEMENT IN AGRICULTURAL MARKETING

Different Types of Organization

Although some definitions were provided in the previous section, these were not especially helpful in distinguishing between the plethora of organizations present in many developing countries.

A four-way categorization is proposed here, based largely on origins and capacity:

1. Northern NGOs with offices in developing countries, usually obtaining funds from donors (including private individuals); this group is quite broad since it encompasses large NGOs such as Oxfam or CARE, with activities in many countries, as well as small NGOs whose activities may be quite focused on a few countries and issues.
2. Indigenous NGOs who have become relatively large, well-organized and able to attract significant funds from international donors and Northern NGOs; sometimes these NGOs were originally created or strengthened by Northern NGOs.
3. Indigenous NGOs that are small, usually focused on a particular geographical area or issue, that obtain small amounts of funding from donors or government, but who struggle to grow or stay afloat.
4. CBOs, membership organizations serving particular interest groups usually in rural communities, whose focus may be broad or quite narrow; these organizations may be formally structured or quite informal; farmers associations, credit groups, and joint marketing societies could all be considered CBOs in the context of this review.

Many countries have laws governing the registration of different types of organizations that may confer a certain tax status or legal standing. Some developing countries have umbrella associations for NGOs. In any particular country it is useful to find out whether an umbrella organization exists, and if so, the types of CBOs and NGOs that tend to be officially registered – recognizing the potential divergence between official requirements and practice.

The Evolving Role of NGOs and CBOs in Development Assistance

In the last 20 years, NGOs have become progressively more involved in development assistance, at every level. The shift from a relief and welfare focus has come about partly in an attempt to address the underlying causes of some of those man-made disasters or to limit the negative consequences of the natural disasters at which they assisted. It has been helped by the increased funding they found they were able to attract. Many Northern NGOs now have policy and research departments, and are a legitimate channel for large amounts of donor funding. At the same time, the role of the state has been redrawn, and in developed and developing countries, there is now a much greater focus on civil society as a way to improve democratic processes and bring about greater accountability in government.

Governments are also seeking ways to be smaller and to sub-contract functions where feasible. Furthermore, funding developing country organizations to carry out development work is considered a way to build indigenous capacity. NGOs working in developing countries have benefited from this trend – either because they are considered part of civil society or because they work closely with many civil society organizations, including CBOs. As Stocker and Barbor-Might state:

- "From the point of view of the donors, civil society was the 'place' where something could be done and, often enough, NGOs were the intermediary institutions or midwives of such remedial programmes [relating to structural adjustment], spanning the gap between donors and CSOs. The funding channels varied, sometimes being directed through Northern NGOs (which might provide 'aid' directly or channel it to one or more partner Southern NGOs or CSOs) and sometimes going as direct funding to Southern NGOs and in some instances even to CSOs. When governments were irredeemably corrupt or oppressive (as, for instance, in Haiti during the Duvalier regime), these programmes seemed to offer virtually the only hope of channelling assistance to the people who most needed it."

This growth in the funding, remit, competencies and responsibilities of NGOs has meant that they have been closely involved in (if not the instigators of) much of the experimentation with practical solutions to pressing problems in rural areas. This is the context in which NGO experiences with agricultural marketing interventions provide a valid and rich focus for this review.

NGOs and CBOs

Although many NGOs share similar altruistic goals, their approaches vary enormously. This is particularly evident in the extent to which they embrace and harness commercial activities to promote broader objectives, or reject this as a legitimate means by which to achieve social objectives. Moreover, amongst those NGOs prepared to use commercial activities as a means to an end, there can be considerable variability in the role these activities are accorded within the development strategy and the competence with which they are planned and undertaken.

Organizations that are Primarily Welfare-oriented

A large number of NGOs and CBOs become involved in agricultural marketing activities, but this is rarely their core business. (There are some notable exceptions amongst some of the international NGOs who have become very experienced in agricultural enterprise, agro-processing and marketing. These include, for instance, TechnoServe, the Intermediate Technology Development Group, Enterprise Works Worldwide and the Cooperative League of the USA.) Many NGOs start with welfare (or social or altruistic) objectives, in areas such as education, health, water, infrastructure and agriculture and gradually shift towards a longer-term development focus. With this shift, small business and income-generation activities take on a greater role. Gibson describes this gradual transformation in terms of a continuum of different actions and attitudes.

Table. NGOs' Evolutionary Path in the Development of small Businesses and Income Generation

From ←	→ To
Relief and welfare	Development
Short-term	Long-term
Ideological	Pragmatic
Community-focused	Individual-focused
Targeted	Self-selecting
Grants	Market interest rates
Amateurish	Professional
Income generation	Small business
Social/ Technical	Economic/ Business
Instinctive	Strategic
Beneficiaries	Clients

Often NGOs and CBOs deliberately work in remote and disadvantaged communities and target the poorest households or individuals. These conditions, in combination with a general relief and welfare orientation, influence the strategies they adopt to achieve their objectives. For instance, direct or indirect subsidies may be used to improve access to markets (*e.g.* through the provision of transport, credit or inputs). An example of a direct subsidy is free or below cost use of transport (calculated on the basis of costs of fuel and driver and perhaps some portion of the vehicle costs). An indirect subsidy might involve charging a commercial (or break-even) rate on the vehicle hire but taking no account of the staff costs of implementing and managing the scheme. Whilst few people would suggest that the subsidy could continue indefinitely, there may be little consideration of how these activities can eventually be shifted to a more sustainable basis. The result is often that the programme attracts participation because of the subsidies, and once it ends there is little enduring impact.

Yet in the short-run these types of activities are attractive to NGOs because they have fairly immediate and visible (if not enduring) impacts and can (with varying degrees of success) be targeted to particularly disadvantaged groups (such as the poorest households, women, refugees, the handicapped or other marginalized social groups). An approach that seeks to use commercial channels may take much longer to develop and may place the intended target group at a disadvantage relative to other members of the community. Even when NGOs and CBOs do not intentionally adopt a non-commercial approach, market-oriented interventions are often subordinate to their core business.

This affects the way they are developed and managed, as well as the way they are perceived within and outside the organization.

Stanton gives four reasons for this lower status:

- Ideological – "...the core work assists those who most need it, income-generating work assists those who will exploit it to the greatest economic advantage";
- Resource allocation – reflecting the ideological perspective;
- Management – "the most willing volunteer for the job rather than external recruitment of people qualified and experienced in business management";
- Finance – these activities may be quite costly to implement and effectively monitor.

Furthermore, marketing activities are often managed and evaluated in the same way as other development activities, with insufficient attention to budgeting and profitability. A survey of income-generating programmes carried out by a range of indigenous and international NGOs in Welaita, south-west Ethiopia, found that little attention was given to the income generated and the profitability of different schemes. However, NGOs using subsidies to target disadvantaged groups would argue that this is a legitimate way to improve the livelihoods of poor individuals, households and communities, particularly in remote areas. Yet even with these subsidies, it may be difficult to have much impact on livelihoods in the most geographically and socio-economically disadvantaged communities. Stanton points out the disadvantages of this type of approach: the frequent failure to make a significant profit, high costs that prevent the NGOs/CBOs from reaching a wide audience, and problems concerning long-term sustainability.

Business-like NGOs and CBOs

- "Organizations which themselves resemble small businesses – in terms of their people, culture, systems, structure and behaviour – are most likely to be successful in encouraging the growth of small businesses."

In recent years, private sector development has increasingly been seen as a viable and important approach to sustainable development. Thus many governments, NGOs and CBOs have focused on the promotion of marketing and small-scale enterprise to encourage greater participation in the commercial sector, as a route to higher incomes, employment generation and growth.

Small enterprise development work, which grew considerably in the 1980s, has contributed to a realization that it is possible to make much greater use of market mechanisms in pursuit of development objectives. This has been shown particularly through the success of micro-credit initiatives, where even very poor individuals are able to repay not only loans but also interest which sometimes covers the costs of providing credit.

Also, social objectives and commercial objectives are not mutually exclusive and many NGOs and CBOs pursue both. The fair-trade movement is a good example of this. Fair-trade organizations use commercial methods to generate social development benefits through improved terms of trade. The important thing is to balance potential marketing success with the social benefit needs of the beneficiaries. Furthermore, selective use of subsidies can still lead to sustainable and successful marketing initiatives, depending on the circumstances, as demonstrated by the CARE Egypt Agricultural Reform Programme. The programme provides information services to smallholder farmers and facilitates linkages to help increase farmer income.

The service is highly subsidized but has proven successful for a number of reasons:

- It helps to link farmers to sources of information outside the programme, thereby fostering the long-term sustainability of relationships and networks;
- Farmers contribute financially, *i.e.* they pay fees for the services;
- The demand for services is farmer-driven and project staff work with farmers to identify production and marketing opportunities.

A similar approach is used by Intermediate Technology in Zimbabwe. Assistance in product development is offered but the initiative must come from an existing business, which must be willing to contribute to the costs of product development (*e.g.* through materials, labour or workshop facilities). The two key advantages of greater commercial orientation and awareness are that cost-recovery enables more people to be reached by such programmes, and sustainability becomes a realistic goal. Gibson argues that NGOs appear to have distinct advantages in pursuing income-generation programmes (''smaller, more flexible, innovative organizations''). His conclusions are firmly rooted in the belief that commercial strategies can serve development objectives:

- "The continuing challenge is to progress from this base so that the economic growth of other developing countries is enhanced, is driven by indigenously owned and indigenously managed enterprises, and reaches the poor and disadvantaged sections of the population.''

Direct Intervention or Facilitation

The marketing role that NGOs and CBOs take on lies somewhere along a continuum between being directly responsible for marketing activities to facilitating beneficiaries/clients to market for themselves.

Direct Marketing Role

The term 'income-generating programme' (IGP) is used to describe a variety of programmes. These range from enterprises owned and managed by the beneficiaries to enterprises owned and managed by the organization which employs the beneficiaries. A number of NGOs/CBOs have established this latter type of small business to generate income to finance their other programmes and reduce donor dependence. CBOs and NGOs can also be more directly responsible for marketing activities. One way of doing this is through out-grower schemes (sometimes referred to as contract farming or satellite production). Such schemes involve smallholder producers providing agricultural raw materials to trading or processing businesses. Often growers work as a group, characterized by String fellow *et al.* as linkage-dependent groups. Generally the marketing arrangements are predetermined: prices or a pricing formula are agreed. They help markets function to the benefit of both producers and the companies or organizations involved.

This type of relationship is beneficial for farmers because they have a secure market for their produce at a predetermined price and the buyer benefits from having a guaranteed source of raw materials and lower transaction costs, which reduces his/her risk and costs. Within the fair-trade arena, the role of, and the marketing channels used by NGOs and CBOs (or alternative trading organizations – ATOs), also varies. Some organizations (such as Oxfam Trading and Traidcraft) take a direct marketing role by acting as wholesalers, with the producers acting as subcontractors producing to order. An advantage of this type of arrangement for producers is that they are guaranteed a volume of sales, thereby minimizing their risk. A disadvantage of this, and of out-grower schemes, is that the producers can be dependent on the trader, and may not have access to alternative buyers or markets if for any reason the trader is no longer able to market their produce.

ADC and Bean Outgrower Schemes in Uganda

The Agri-business Development Centre (ADC) in Uganda helped to establish bean outgrower schemes in Kasese and Kibaale districts together with the Uganda National Farmers Association (UNFA) and Bugangaizi Export Commodities Limited (BEC). The purpose of the schemes was to integrate poor rural farmers into the bean market, through marketing agencies, to provide them with an additional income source. ADC's role was facilitating the supply of bean seed and training some of the producers as 'farmer-extensionists' to provide follow-up extension advice to other producers. UNFA and BEC promoted and implemented the programmes. They were responsible for managing seed supply, training and extension and organized marketing (procurement and collection) of the crop.

The outgrower schemes have been successful from the producers' perspective in several ways:

- The numbers of farmers reached has steadily increased;
- Output levels have increased (participation, acreage and yields have all increased);
- An effective farmer-extension system has been created, and the adoption of new varieties and improved production methods has been good;

- Planting materials have been maintained and expanded through seed multiplication;
- Household income from bean sales has increased.

From the trader perspective, a strong linkage was formed with a private buyer in Kasese, who procured beans through UNFA and sold to a Kampala-based exporter. In Kibaale the linkage was weaker, due to the inability of BEC to raise finance. This lack of capital was a key constraint, and undermined the efforts that had gone into developing the scheme and building effective outgrower loyalty. Also, competition from other traders emerged for the beans, which affected potential profit. This highlighted the importance of product selection within outgrower schemes and the need to consider diversion factors (whether the product can be used or sold outside the outgrower scheme) and the buyer's (financial) exposure ratio (the likely cost of obtaining the crop against anticipated sales value).

Facilitative Role

Other NGOs and CBOs play a more facilitative role. They assist individuals, groups and communities to market for themselves. This includes both improving access to, and benefits generated from, existing products and existing markets as well as creating new products and new markets (*e.g.* through technology development and processing). There are a variety of ways in which organizations facilitate marketing, including: strengthening the capacity of individuals, groups or communities (through group strengthening and training); developing linkages to traders and other stakeholders in the marketing chain (*e.g.* input suppliers, credit sources and transport agents); and linking farmers to relevant market information. This type of facilitative role is beneficial for a number of reasons: being less interventionist, it is likely to generate more sustainable marketing activities and linkages; it is likely to be achieved at lower cost than if the NGO was more directly responsible for marketing activities; and, therefore, it facilitates reaching a wider audience.

TYPES OF MARKETING INTERVENTION

As indicated in the previous section, there are many different ways in which NGOs or CBOs may intervene to improve access to agricultural markets.

In this section interventions are discussed in eight non-exclusive categories that describe aspects of the intervention strategy:

1. Intended beneficiaries
2. Skills and training
3. Access to agricultural inputs

4. Agro-processing technologies
5. Marketing linkages
6. Credit programmes
7. Marketing information
8. Holistic approaches

Any particular marketing intervention may comprise elements from several categories (*e.g.* inputs and training, or technology, training and finance). The concepts and experiences associated with each category are reviewed, permitting some preliminary conclusions on the more promising strategies.

Intended Beneficiaries

Individuals, Groups or Communities

Different NGO and CBO marketing initiatives operate with different beneficiary or client structures: some work with whole communities, some with groups and others with households/individuals. The choice of appropriate structure will depend on a number of factors. Research carried out by the Plunkett Foundation and experience of CARE's Development through Conservation project in Uganda suggests that working with village associations or whole communities is more difficult than working with smaller groups. The latter are more focused, more specialized and more likely to have a common goal. Many NGOs and CBOs implement their marketing interventions with groups or associations. Not only are there advantages to the NGOs and CBOs of working with groups, but there may be advantages for the farmers themselves of marketing collectively. The groups can be existing groups (*e.g.* women's groups, savings and credit groups, social groups and so on) or newly formed groups.

The potential advantages of farmers collectively addressing marketing constraints include:

- Economies of scale, through joint purchasing of inputs and joint marketing of products;
- Improved access to finance, where credit organizations favour group loans, or where pooled resources provide the necessary down payment; this can overcome problems of larger investment needed in, for example, processing technologies, storage facilities or transport;
- Collective bargaining power;
- Lower transaction costs (for producers and traders).

Yet there is much evidence that, in general, small enterprises owned and managed by individuals are more successful than group enterprises.

Technology adoption in particular is felt to be better amongst individuals than groups, although groups can be an appropriate vehicle for technology transfer if the technologies are subsequently employed by individuals. This is echoed by String fellow *et al.* who argue that group enterprises are more likely to succeed when based on joint marketing rather than joint management/ownership of assets, because the latter requires more complex skills and experience.

It has also been found that external organization and management of groups can prevent the development of entrepreneurial skills. There are no definitive rules on which structure is appropriate. It will depend on the type of intervention and the objectives of the individuals and NGOs/CBOs. On the one hand, NGOs need to remain open to working with groups where appropriate, recognizing the potential to build capacity, reduce transaction costs, and introduce activities with a higher investment threshold. However, an understanding of the reasons behind group formation is important and care should be taken not to over-burden groups formed for social rather than economic reasons. Sometimes individuals, often women and particularly in rural areas, prefer to work together. Furthermore, commonly cited problems attributed to groups per se have been found to be less important when there is strong group leadership and cohesion, and when there is good group organization before any external intervention takes place.

String fellow *et al.* point out that a non-interventionist approach – letting producers decide for themselves whether they operate as individuals or groups – allows individuals to develop appropriate structures to build necessary skills and solve their own problems. People will not work well within an imposed structure. As Gibson succinctly states: "Enterprises working in a market environment (whether collectively owned or individually owned) have ultimately to make a surplus in order to survive. The 'bottom line' for NGOs is to support the structures which will work best according to this criterion."

Rural Poor

The majority of NGOs and CBOs deliberately target their activities at poor households and poor communities. Those that focus on agricultural production, processing and marketing are often found in remote rural areas. This focus has implications for the approach adopted by the NGOs/CBOs and the marketing initiatives they develop. Whereas the trend in marketing approach adopted by NGOs has moved towards a more business-oriented, facilitative approach in recent years, some authors argue that this is less likely to succeed in remote rural areas than in less remote, higher-potential areas. Moreover, within rural communities, some poor individuals and groups are considered too remote and disadvantaged to be able to benefit from marketing interventions.

The most vulnerable households and women-headed households in rural areas are particularly disadvantaged; they tend to have more restricted access to information and services and tend to be more risk-averse than other households.

Some households may only benefit indirectly from a marketing intervention through any impact it has on the labour market. Consideration should also be given to the types of intervention most appropriate to a particularly disadvantaged target group, particularly where the transfer of hardware and entrepreneurial skills are envisaged. For example, food processing technologies are often predominantly adopted by the rural elite, due to their more developed entrepreneurial skills and access to finance. This may nonetheless reduce poverty, through improved markets and employment generation, but the strength of these linkages varies, making it difficult to generalize. Certainly more information is needed on the extent to which the poor benefit directly and indirectly from NGO and CBO marketing interventions, and about which approaches and types of intervention are having most impact.

Women In common with other organizations, NGOs and CBOs often find it very difficult to effectively target benefits to women. Yet women are frequently amongst the poorest members of rural communities and face particular constraints in improving their livelihoods. Moreover, their traditional role in food crop production and (sometimes) domestic marketing, as well as the influence they have over child welfare and nutrition, makes them an obvious target for poverty-focused agricultural marketing interventions. In their review of women's role in post-harvest operations, Gordon *et al.* conclude that:

"Where interventions are intended to benefit the poorest women, attention should be focused on particular issues:

- The needs of female-headed households, which feature disproportionately amongst the poor;
- Crops and processes used in marginal areas;
- Carrying fuel and water, because so many women are affected;
- How poor women earn income – so that new technology really does benefit them;
- Understanding the broader processes which determine how benefits are distributed;
- Household level and informal sector activities, where the poorest people earn their living."

Goodland *et al.* make a number of suggestions on how women's participation in rural finance programmes can be increased, the spirit of which is equally relevant here:

- Out-of-hours opening or mobile services, or locating services in places women frequent, for example, marketplaces;
- Relaxing literacy requirements;
- Flexible collateral requirements, for example, accepting jewellery rather than land;
- Allowing loans of a suitable size (usually small).

Skills and Training

The types of training offered by NGOs in support of marketing initiatives can include group strengthening, general extension, marketing and specialized training.

Group Formation and Strengthening

The advantages of working with groups have already been highlighted. Despite these, group establishment and operation have generally proven more difficult in practice than expected. Some organizations undertake group strengthening and training (*e.g.* in recruitment and management, structure and leadership, financial and business management skills and so on) before marketing activities are initiated, or as part of the marketing intervention.

The case of oil palm processing with the NGO TechnoServe is a good example. The danger of not assessing the performance of existing groups and providing any necessary training in group strengthening is that marketing activities implemented with these groups can fail due to general group weaknesses, rather than problems with the marketing activities per se. Stringfellow *et al.* studied farmer co-operative enterprises and their findings highlight the importance of not over-estimating group capacities, and the need for long-term involvement in building group capacities. Often groups fail because they have been formed too quickly and too much is expected of them.

They also found that group enterprises are more likely to succeed when based on joint marketing rather than joint management/ ownership of assets which requires more complex skills and experience. It is also important to consider the most appropriate group size and whether there is a culture of working as a group. The Co-operative League of the USA (CLUSA) working with CARE in Mozambique has developed a thorough approach to the development of farmers' associations. They have attempted to make this approach more sustainable by developing the capacity of a local NGO to take-up and continue these activities, and by promoting a structure within the farmers' associations that permits ongoing development. The approach has been criticized for being too costly, but the results to date, albeit over a short duration, are nonetheless very impressive. There is now a need to critically evaluate this approach, to identify the direct and indirect subsidies provided by the NGOs, and assess which components have the most prospects of sustainability.

TechnoServe and Oil Palm Processing in Ghana

TechnoServe is an NGO whose activities focus on the provision of food processing technologies for rural communities. The initial focus of its work was small-scale oil palm processing and extraction in Ghana, but it has subsequently expanded to include processing service centres, inventory credit schemes and production and processing of non-traditional export crops. The purpose of the oil palm scheme was to build on traditional processing (which was laborious and slow) and to exploit the potential for expansion in a strong market, by introducing small-scale mechanized oil mills to community-based groups. The approach adopted by TechnoServe went beyond providing the technology and technical assistance. It was recognized that entrepreneurial skills in the communities were weak, and that group development activities and financial and business training (including linkages to formal credit and extension services) were necessary. This integrated training and support package clearly contributed to the sustainable adoption and management of the oil processing enterprises. However, the Techno Serve approach has also been criticized. The training and support provided to groups are heavily subsidized. Case studies of the oil palm enterprises indicate that the skills are held by only a small number of active group members, and that because of social norms, there is a reluctance to pass these on. A more serious criticism of the approach is that although it has been successful in transferring skills specific to the oil palm enterprise, it has not managed to create more general entrepreneurial skills and characteristics amongst the group members.

Training and Extension

The starting point of a number of CBO/NGO marketing initiatives is production. This is because to market successfully, farmers need to produce and sell what is in demand, at a profit. Often existing markets could be accessible to farmers (either on their own or through linkages with traders), but marketing is constrained by the low volumes or poor quality of farmers' crops. Improved production practices are important for increasing yields of existing crops, new varieties and new crops. Yet government extension services are often lacking or extremely under resourced. NGOs, therefore, often assume a role in providing, or facilitating the provision of, relevant extension information. This may be through strengthening existing extension services or through the establishment of alternative services, such as farmer-extension schemes.

When the CARE Egypt Agricultural Reform Programme began, it focused on improving crop and livestock production. Now, however, the project has become more market oriented and requires farmers to examine the market for their products prior to improving production. Other NGOs and CBOs provide marketing training to groups and individuals. This includes training in production and marketing systems, constraints and opportunities, market demands (products and service) and how to assess whether products can be supplied profitably. The benefits of providing producers with this type of training are that it develops their capacity to analyse markets for themselves and, therefore, allows them to respond to changing market opportunities and threats. They are not dependent on the NGO/CBO to identify marketing opportunities for them in the longer term. It is also important for NGOs and CBOs not to create parallel services and for different development organizations and governmental bodies to co-ordinate the training and services they provide. In Uganda, for example, a task force was established to prepare guidelines on how to integrate NGO activities into district agricultural programmes. Specialized or vocational training is usually provided to individuals or groups when the marketing intervention relates to a new product or new technology, and requires new skills. Examples include production targeted to a new niche market or particularly quality-conscious export market, or training in the use of an agro-processing technology, such as an oil press. Increasingly organizations which provide skills and training for enterprise development are charging trainees a fee. This helps NGOs and CBOs with limited funding to reach a wider audience. More significantly, it has been found that charging a fee increases the proportion of trainees who actually make effective use of their training. The entrepreneur, making the decision to invest money and time in training, is in effect making a risk assessment.

Improving Access to Agricultural Inputs

Poor access to inputs directly influences the level and quality of production. Even in the poorest parts of Africa, there is still demand for farm implements and good quality seed. Less-poor farmers may make selective use also of fertilizer and pesticides. Input subsidies are a particularly vexed issue. Some argue that they are needed to provide a short-run boost to production and incomes. Yet they are also disruptive and undermining of sustainable commercial development. At a workshop in

Uganda, participants highlighted the negative effect of farm input relief programmes in neighbouring countries on the development of commercial input supply networks in Uganda. In Malawi, the starter-pack scheme (distribution of free seed and fertilizer) implemented in 1999 and 2000 has boosted production there, but it has also deprived other low-income farmers in neighbouring countries (notably Mozambique) of a traditional outlet for their surplus production. Gordon discusses five sets of issues affecting access to inputs: affordability; availability; information; risk and uncertainty; and the overall commercial context. Although credit is often assumed to hold the key to improved access, other ways to improve affordability are also identified: timing input sales to coincide with times when farmers have cash; selling inputs (*e.g.* seed) in small pack sizes suited to small producers; and lowering prices, by making cost reductions in distribution and marketing (*e.g.* through bulk purchases, transport sharing arrangements, and farmers' groups taking on more responsibilities).

Whilst NGOs may play a role in providing credit or promoting some of these other strategies, a role for CBOs (*i.e.* farmers associations) is much more apparent. In countries that have succeeded in significantly improving access to inputs, farmers' associations have acted as a key vehicle for input distribution. Many consider the physical availability of inputs to be a more important constraint to access, with thin and unreliable rural distribution networks in most African countries. A recent development involves innovative approaches to the promotion of input stockist networks by NGOs, illustrating what can be achieved through constructive partnerships between the commercial, private non-profit, farming community and government sectors. These initiatives typically involve training stockists and may involve loans or loan guarantees.

There is also a growing interest in ways to improve and build on traditional informal seed systems. Information constraints are also important – be they in terms of information gaps (basic research on fertilizer response, for instance) or information flows. Although farming is inherently risky, better information reduces uncertainty, enabling farmers to make more informed production decisions. Gordon concludes with some general observations, which also have relevance to the activities of NGOs: "In addition to policies aimed towards the general development of rural economies, a number of more specific policy recommendations are made: avoid actions which undermine the development of sustainable commercial input supply networks; support input markets by setting standards and regulations, and providing information and training; promote synergistic partnerships between commercial, private non-profit, farming community and government sectors; and fill critical research and information gaps."

Agro-Processing Technology

Small-scale farming households in remote rural communities generally find that they operate in markets comprising many producers of undifferentiated products, which leads to price competition and low profit margins. Access to processing technology can provide new market opportunities by reducing perishability or adding value in other ways. Processing technologies can range in scale from household-level 'lowtech' processing to fully mechanized factories. Household level processing has two main functions. It can add value and it can preserve the product, thereby increasing the time available for marketing. Other advantages of small-scale agro-processing enterprises are that they can create employment at low levels of investment that make effective use of local resources. Enterprises owned and managed by individuals or households are often more successful than group enterprises, so technology development organizations need to be aware of this need for small-scale technologies. However, many NGOs and CBOs work with farmer groups either because it is the structure that the farmers prefer (there may be a culture of group activity) or because the NGO/CBO prefers this approach (due to financial and coverage reasons). Yet group approaches to the adoption of agroprocessing technology are often weak and entrepreneurial skills are less evident than when working with active individuals. In these circumstances, careful group selection is required, as well as consideration of ways in which the need for entrepreneurial skills can be reduced by introducing a third party, such as a private company offering marketing services to small-scale processors. It is necessary to ensure that there is market demand for the technologies and/or their end-products and that the technologies are appropriate (*e.g.* taking into account gender issues), rather than developing technologies for their own sake. It is also important that the wider enterprise environment is considered and that institutional arrangements enable smallholder access to markets. Also, as will be examined further, many marketing constraints faced by farmers are interrelated. In terms of technology development, other marketing factors such as access to credit, adequate managerial and technical skills, and market information all influence technology choice decisions.

The Ram Press – Oilseed Processing Technology Introduced by ATI

In the 1980s, ATI (now Enterprise Works Worldwide) started developing a manual ram press (also called the Bielenberg press after the engineer who developed the prototype) to produce edible oil in remote parts of Africa without electricity and with poor access to markets. The first presses were intended for soft-shelled sunflower seed in Tanzania, but the model has since been adapted for use with other oilseeds, including coconut and sesame. The first presses were arduous to use but later models improved on this and could be operated easily by one person. It was hoped that they would improve nutrition by improving access to energy-rich food, whilst also increasing farmer incomes and creating enterprise and employment in rural areas. Indeed, local entrepreneurs involved in small-scale oilseed processing can earn two to three times more gross income, compared with selling the seed to processing factories. In Tanzania, costs could be recovered in just one 3-month season. ATI worked up the promotion methodology over many years in many African countries. The ram press is quite simple and is supplied with a filtration device and tools for maintenance. Training, information and support on proper use and socio-economic and nutritional benefits of an oilseed processing operation are important elements of the extension programme adopted by ATI and its partners. In Tanzania, ATI sold the presses on credit for many years. Elsewhere, whilst varying in their systems and repayment rigour, credit has been an important factor affecting uptake. Depending on the model and manufacturer, the presses cost between US$ 150 and US$ 300. ATI has been innovative in promoting the commercialization of small-scale oil processing in Africa with its Regional Oils Programme. Its focus shifted from providing technical assistance to NGOs involved in small-scale oilseed processing to a private sector approach that emphasizes commercialization and mass manufacturing of ram presses. As a result, private enterprises have been developed, given assistance with market surveys and business plans, and are becoming responsible for the manufacture and wholesale of ram presses. ATI also stress the importance of the participatory processes used that involve economic actors in the community – owners and workers, press manufacturers, sales agents and oil-consuming families.

Marketing Linkages

Marketing problems identified by producers are often attributed to the commercial sector and the capacity to access it: lack of buyers, unreasonably low farm-gate prices, inflexible requirements, and so on. Links between NGOs/CBOs, the governmental sector and commercial agents are usually weak. Yet working together with the private sector is an important way for farmers to access relevant market information, technologies and new market opportunities. It is also a way for NGOs and CBOs to reach a wider audience with limited funds. As Kleih has stated, although NGOs are making a very positive contribution to rural development, they generally only reach around 1–2% of farming households in any one country. Private sector agents are sometimes willing to collaborate with NGOs and farmer groups to share the costs of providing training and information, if they see it as an investment through which they can increase their own revenue. In these circumstances, both the producers and traders can benefit. Developing and strengthening relationships between farmers and traders can reduce transaction costs, transport costs and risk on both sides. It was noted above that approaches that use and strengthen existing private sector production and marketing channels, rather than seeking to override them or invent new ones, have been particularly effective in addressing marketing constraints and in sustaining linkages beyond the life of the NGO marketing project. The Smallholder Agribusiness Development Promgramme (SADP) in Malawi, established by the American Co-operative Development Initiative and Volunteers in Co-operative Assistance (ACDI/VOCA), links farmers to existing private-sector markets, rather than establishing new market channels. This has contributed to its success in working with large numbers of farmers and achieving sustainability. CLUSA and CARE have also focused on developing linkages between farmers and traders in Mozambique.

Credit Programmes

Credit merits particular attention in the context of agricultural marketing and processing. Farmers groups and NGOs often recognize a lack of credit as a critical constraint to the development of new initiatives and many seek to remedy this through credit interventions. The aspects discussed here are drawn from a recent synopsis by Gordon.

Rural Finance

Rural finance comprises credit, savings and insurance (or insurance substitutes) in rural areas, whether provided through formal or informal mechanisms. The word 'credit' tends to be associated with enterprise development, whereas rural finance also includes savings and insurance mechanisms used by the poor to protect and stabilize their families and livelihoods (not just their businesses). Rural finance comprises informal and formal sectors. Examples of formal sources of credit include: banks, projects and contract farmer schemes. Reference is often made to micro-credit. Micro underlines the small loan size normally associated with the borrowing requirements of poor rural populations, and micro-credit schemes use specially developed pro-poor lending methodologies. Rural populations, however, are much more dependent on informal sources of finance (including loans from family or friends, moneylenders, and rotating or accumulating savings and credit associations). There is an enormous literature on rural finance and micro-credit, much of which relates to small-scale enterprise. (Many micro-credit schemes specifically exclude agricultural production activities, even in rural areas, because they are considered high risk.) The discussion here focuses on particular aspects of rural finance that are relevant to NGO and CBO agricultural marketing initiatives. First, typical features of micro-credit schemes, which tend to be run by NGOs and often work with community groups, are reviewed. Attention is then focused on two types of credit scheme with particular relevance to agricultural marketing interventions: inventory credit – a relatively new departure in smallholder agriculture, which is attracting increasing attention from NGOs and private banks; and outgrower schemes, which tend to be supported by the private sector but often work through producer groups. Some NGOs support other 'hybrid' rural finance initiatives that may affect agricultural marketing. These include stamp-based savings groups and rotating savings and credit associations (ROSCAs), where pooled savings may facilitate access to additional loans, and permit crop purchase/assembly/marketing activities or improved access to agricultural inputs.

Micro-credit and the Rural Poor: The Issues

Rural credit would not be the focus of so much development effort were it not for widespread market failure (*i.e.* failure to provide the services people want) in rural financial services in developing countries.

Reasons for market failure include:

- The lender does not know the default risk of each potential borrower and to collect this information is costly;
- It is costly to ensure that the potential borrowers take those actions which make loan repayment more likely;
- It is difficult and costly to enforce repayment;
- The cost of providing services to the rural poor is high because they are located in remote areas, want to borrow small amounts, and illiteracy, lack of experience of banks, and lack of collateral necessitate the development of tailored approaches.

What are the implications of this for agricultural activities? Firstly, all these types of market failure apply to agricultural lending in developing countries. Lack of information on the risk of default is particularly germane to agricultural enterprise. Farmers do not keep records, so it is difficult for them to produce the information that might convince a bank of their creditworthiness.

Rural market transactions are largely informal, so it is difficult for the bank to collect independent information on prices. Farming is clearly a risky business because of weather, pests and market fluctuations and it is difficult for a bank to assess the degree of risk associated with particular activities. The rural poor do not have a track record, or referees who will vouch for their competence and reliability. Making sure that farmers keep to their business plan, using loans as intended, and carrying out tasks to schedule, is also costly – although this might make loan repayment more likely. Enforcing repayment is also difficult. This requires monitors who know when crops are sold, or agreements with merchants to pay the farmer net of what she/he owes the bank, or effective penalties such as seizure of assets or prosecution. Farmers rarely have collateral acceptable to banks. They may not have clear title to the land they farm, or even if they do, rural land markets may not function well enough for land to be considered a 'bankable' asset. Poor farmers, moreover, rarely have other bankable assets. They might own a bicycle, and have a store half full of grain, but were a bank to seize such assets the cost of doing so would probably exceed their sale value.

The poorer the farmer, the fewer are his/her chances of borrowing from the formal sector. Women, particularly poor women, face even more problems in obtaining credit. Land title, where it exists, is usually held by men. Women often have little control over other factors of production, particularly for the 'bankable' cash-cropping activities. In some countries women may only borrow in the names of their husbands, if at all, and literacy rates for poor women are almost always lower than those for men. The irony is that numerous studies show that women tend to repay loans more reliably than men. Numerous projects, government schemes and NGOs engage in loan programmes targeted at the poor. Some of these work well, but many are unsustainable because of high and subsidized costs, and high rates of default. Moreover, many miss their target, with the benefits captured by the less poor. The poor depend overwhelmingly on the informal sector. Micro-finance is the response to market failure in 'conventional' banking services for the rural poor. It responds to the needs of low-income households.

Sound schemes tend to be characterized by:

- Small, short-term loans, and savings mechanisms;
- Simplified loan appraisal procedures;
- Innovative approaches to collateral;
- Rapid approval/disbursement of repeat loans after repayment;
- High transaction costs;
- High repayment rates;
- Savings and loan services provided at a location and time convenient to the poor.

Thus, micro-credit schemes are often associated with: group-lending (where peer pressure effectively substitutes for collateral, and other group members may take action to prevent one member defaulting, for instance, by providing labour to assure timely harvest); extension inputs arranged by the micro-finance institute (MFI); and mobile banking arrangements. Cashflow analysis may concentrate on overall ability to repay the loan rather than a particular investment project. In some respects, MFIs try to imitate the strengths of the informal sector (using local information to ensure repayment, for instance) and some MFIs are experimenting with ways to link their operations with some of the informal sector financial agents.

Agricultural marketing loans are not common in MFI loan portfolios, but some micro-credit schemes do provide marketing finance or input loans. MFIs can include government and commercial banks, NGOs and savings and loan co-operatives. Some of the largest MFIs are Asian (for instance, Grameen Bank in Bangladesh has 2 million borrowers and savers, Bank Rakyat Indonesia/Unit Desa system has 2 million borrowers and 12 million savers, and the Bank for Agriculture and Agricultural Co-operatives (BAAC) in Thailand has 2.5 million clients). Coverage is less in Africa, but still growing.

K-Rep in Kenya has 15000 clients and the Fe´de´ ration des Caisses d´ e´pargne et de Cre´ dit Agricole Mutuel (FECECAM) credit movement in Benin has 200 000 clients. Whilst these figures are impressive, they indicate a major difficulty: there are too few services to go round – and still fewer likely to become sustainable. Many schemes find it difficult to graduate from subsidized operations to full cost-recovery. Uncosted but critical inputs by NGO staff are commonplace. Interest rates are often set below market rates (even at negative real rates, where high inflation prevails). Too little attention may be focused on repayment and follow-up; and farmer experience of subsidies, free inputs and loan amnesties may all contribute to low repayment rates.

Contract Farmer Schemes

Contract farmer or outgrower schemes focus on very specific needs. They usually operate in situations where a processor or trader faces a supply constraint and, therefore, wishes to promote production of crop x, and has access to sufficient resources (own or loaned) to do so. Inputs are usually provided in-kind (to reduce diversion to other activities), accompanied by extension, and the cost of these services is recouped out of the price paid to the farmer when the crop is harvested.

The degree of supervision and type of inputs provided varies greatly. High value horticultural products produced for export may be closely supervised, for instance, whilst supervision of a cotton crop (which still has to undergo considerable processing before it reaches the consumer) may be minimal. Viable schemes must include mechanisms to minimize on-farm consumption or side-selling (farmers selling their crop to an alternative buyer and, therefore, avoiding loan repayment), such as:

- Sharing of information and co-operation amongst crop buyers;
- Effective penalties against farmers who default (exclusion from future schemes, seizure of assets, prosecution);
- Peer pressure achieved through group lending.

Some of the longer-established cotton schemes in West Africa are wide ranging in scope. Considerable long-term effort has gone into capacity building of farmers' groups. This enables farmers to take on a greater share of tasks in input supply and crop assembly, which both reduces cotton company costs and increases cash revenues to farmers. Inputs and extension are available for other crops in the farming system (not just for cotton) effectively reinforcing and expanding the benefits accruing from cotton production.

These types of schemes are common for smallholder annuals grown for export such as tobacco, cotton and horticultural production, where the on-farm investment costs are relatively low, and the pay-back period short. (Thus, greenhouse production of flowers for export in Zimbabwe is not a smallholder activity.) In Africa, the cotton schemes are undoubtedly the largest.

About 300000 Ugandan farmers benefit from an input scheme organized by the cotton ginners. In Zimbabwe, around 60000 communal farmers take input loans from cotton companies (and many more participate in other schemes intended to increase cotton output). In Mali, around 100000 rural households participate in the cotton input schemes. The schemes in Mali and Zimbabwe both depend on a network of strong farmers groups, and the Ugandan scheme is likely to develop in that direction as it matures, seeking less costly ways to reach its target group.

Crop Market Characteristics

If there is limited on-farm consumption and local marketing, it may be possible to collect loan repayments at the point of sale. However, unless there is a crop purchase monopoly, or agreement to share information between different buyers, the farmer may be able to avoid repayment by 'side-selling' (selling to another buyer).

Input qualities

Inputs are often provided in-kind to reduce 'diversion' of the input away from the targeted crop. Diversion is lessened if there is a limited alternative use or market for the inputs, or if (unusually) returns to use of the input are greatest for the crop in question.

Commercial/Credit Context

Prospects for viable operation of the scheme are greater if farmers treat farming as a business and are integrated into markets, and if there are supportive legal/political/ contract enforcement institutions. A recent history of loan amnesties, default without penalty, and subsidized inputs may undermine the operation of viable schemes.

Modus operandi of scheme – best practice:

1. Group schemes for peer pressure.
2. Group or individual schemes backed up by monitoring/good information, support staff, and ability to act.
3. Incentives for repayment and penalties for non-repayment.
4. Appropriate incentives for field monitors/co-ordinators.
5. Training provided to farmers extension and business management.
6. Developing relationship/trust/loyalty through field presence/ contact.
7. Accessibility of scheme – minimize red tape and transaction costs; organize so that the location and timing of contact is convenient to farmers.
8. Effective and timely monitoring of input use and crop marketing.

Inventory Credit

Another mechanism for financing agricultural trade exists in inventory credit or warehouse receipt systems. This involves a tripartite agreement between a bank, a borrower (usually a trader), and a warehouse operator. Essentially, a trader is able to use an existing stock of, say, grain as collateral for a bank loan providing certain conditions are met. The grain must meet certain verifiable specifications and be stored in a warehouse operated by a third party. The bank will usually lend up to a certain percentage (perhaps 80%) of the value of the grain at that time (at harvest time when prices are low). The trader can then use the loan to acquire additional stocks of grain, and is usually required to repay the loan during the 'lean' season before grain prices start to fall once more with the next season's harvest. The trader must settle all the warehousing costs before she/he removes the grain, and is generally required to repay the loan at this stage also. In the event that she/he is unable to repay the loan, the bank can seize the grain and sell it.

There is a lot of interest in inventory credit at the present time because of the scope it offers to provide traders with capital to fuel agricultural marketing. However, its use is limited by a number of factors: it can only be used for crops that are relatively non-perishable, with reasonably predictable pricing scenarios; by definition it is targeted to larger, more sophisticated traders, able to purchase initial stocks and fairly confident in their negotiations with banks and warehouse operators; and it is critically dependent on the necessary institutional infrastructure (bank willingness to lend against inventories, warehouse systems which can operate to the standards and within the necessary legal framework, and supportive and enforceable legal institutions). NGO involvement in these schemes can take a number of forms. It may include a role as an MFI providing the credit, or it may include working with farmers' groups who effectively become the 'trader' in the tripartite agreement, buying crops from other farmers (or their membership), for storage and sale later in the season when prices are higher.

Who Benefits from Rural Credit Schemes?

Particularly those of the poor, tend to be barred from formal sources of credit. Lack of capital is a widely recognized constraint to marketing and enterprise development. Both producers and traders need capital. It is needed to purchase inputs, to allow farmers greater flexibility in the timing of sales and to allow traders to engage in intra-seasonal storage. In recent years there has been significant growth in the provision of micro-credit services to help businesses develop. Although there has been reasonable success in terms of widespread uptake (particularly amongst women) and high loan-recovery rates, the impact has been relatively disappointing.

Table. Who and what is Excluded or Less Excluded from Rural Credit

Tending to be excluded ←	**→ Tending to be less excluded**
Formal sources of credit	
Women	Men
The poor	Less poor
Illiterate	Literate
Agricultural activities	Non-farm enterprise
Landless	Those with land
Those with no bankable assets	Those with other collateral
Food crops	Cash crops – especially export crops
Unforeseen urgent needs	Planned investments
Informal sources of credit	
Long-term	Short-term
Large loan size	Small loan size
Seasonal inputs and investment	Crisis loans

One of the reasons for this is that much of the credit has been used for basic trading and simple processing activities, which provide limited scope for increased productivity and added value. There has also been relatively little use of credit for technology development. This has led to an expansion of similar enterprises, producing a limited number of goods and services for an already saturated market. A second reason is that small producers and small enterprises cannot afford to take the risk of developing, testing and promoting a new product. They tend to be risk-averse because of limited financial resources and because they lack access to the necessary information to help them profitably change their businesses.

Marketing Information

Market information generally refers to market price information, and in some cases includes information on quantities. Marketing information is a wider concept, including information on marketing channels, buyers, quality standards and so on. Post-independence, many market information systems (MIS) in developing countries were government-run but most are now either defunct or confronting major problems. Yet accurate, appropriate and timely marketing information (on prices and on marketing issues more broadly) is very important for producers and traders. Market information informs production, processing, storage and marketing decisions, facilitates the spatial and temporal distribution of products and can strengthen the bargaining power of producers.

Yet in practice, MIS have repeatedly proven to be unsustainable or have failed to provide timely and useful services. This has led some authors to argue that users should pay for market information. However, others argue that in the context of small-scale, resource-poor farmers in remote communities, payment for marketing information is unrealistic and information should continue to be considered as a public good. This is particularly true for information provided using mass-media, which makes cost-recovery very difficult.

Also, farmers in remote areas with poor coverage by traders often point out that knowing market prices does little to improve their bargaining position if there is little competition amongst traders. The MIS in Mali, which was reorganized in 1998, illustrates some more positive features. The objectives of this reorganization were to ''create a decentralized MIS which is efficient, viable and sustainable''. A participatory needs assessment was carried out initially to identify the marketing information needs of different stakeholders.

Private sector stakeholder involvement and their needs are recognized as one of the key features of the MIS. One of the objectives of this MIS is to recover some of the costs through the sale of information to commercial stakeholders. However, for remote areas, the information is viewed as a public good, and it is not anticipated that the system will be able to work without donor support. NGOs may have a role to play in collecting and/or disseminating market information where local government or private sector capacity is weak. It can involve assistance to local government in establishing MIS, training information officers and/or provision of transport and equipment.

However, it is important that the institutional side of any established MIS is sustainable; the creation of new and unsustainable units, which do not form part of the local institutional set-up, should be avoided. Also, it is important that MIS are co-ordinated. The different marketing information requirements of producers and other stakeholders in the marketing chain may require several different sources of information (*e.g.* NGOs and CBOs, government extension and research services, private companies and so on). However, co-ordination is required to avoid duplication and to use scare resources efficiently.

Holistic Approaches

Numerous authors have indicated that marketing constraints cannot be tackled individually. The experience of many NGOs and CBOs has shown that isolated marketing interventions, for example, the provision of new processing technology or market information, are unlikely to succeed. A holistic approach is needed to marketing and enterprise development, to look at the whole range of marketing constraints so as to improve the terms on which farmers participate in the market. Indeed, this is the strength of the private sector; it adopts a holistic approach when assessing the feasibility of enterprises and balances market demands, technology, infrastructure, training and other requirements. Many of the marketing interventions already described in this review have been successful through adopting a holistic approach. For example, the Appropriate Technology oil palm press initiative involved much more than just design and provision of the technology, likewise the CLUSA approach to farmers associations. Training was provided to recipient groups in production, marketing, technical aspects of the oil press including repairs and agribusiness management, and linkages with manufacturers and traders were facilitated. The Smallholder Agribusiness Development Programme (SADP) in Malawi is another good example of a holistic marketing programme.

ACDI/VOCA – Addressing a Range of Marketing Constraints in Malawi

The Smallholder Agribusiness Development Programme (SADP) was established by the American Co-operative Development Initiative and

Volunteers in Co-operative Assistance (ACDI/VOCA) and its primary aim is to improve market access for farmers. The programme was initiated to work with tobacco farmers (the largest and most profitable cash-crop sector in the country) at a time when the sector was being deregulated in the early 1990s. The programme works with existing clubs or associations of tobacco farmers, who are responsible for bringing farmers together to jointly solve their problems. The programme deliberately selects stronger functioning farmer clubs to work with and they are expected to operate as commercial businesses. The services provided by the programme are broad and address a range of marketing constraints.

They include:

- Capacity building of groups and associations (through management and financial training);
- Assistance in identifying problems;
- Assistance in identifying and developing solutions by:
  - Providing training and market information
  - Facilitating linkages, identifying private sector service providers
  - Negotiating transportation and other service agreements
  - Developing improved delivery systems.

The programme strives to be sustainable through a number of measures:

- Farmers contribute financially through club association membership fees;
- Clubs and associations do not provide services directly (this would need more capital and management training) and are linked to commercial credit sources, rather than receiving grants;
- Where possible, services are contracted out to the private sector;
- Establishing a national organization (NASFAM) to provide these services beyond the life of the programme.

A number of the marketing services have proven profitable and sustainable (11 of the 12 farmer associations were financially sustainable in 1998). Those that have not been include the collection and dissemination of marketing information, auditing and monitoring performance of farmer associations and helping farmers to diversify their crop base.

Conclusion

The argument in favour of agricultural and food enterprises, in developing countries, becoming more customer oriented is a compelling one. Average incomes in developing countries are low and so the marketing systems which deliver agricultural and food products have to be efficient if they are to deliver food and other products at affordable prices. Moreover, when a country does experience economic growth this is normally accompanied by an acceleration in the rate of urbanisation. The end result is that greater demands are placed upon farmers. Marketing systems have to be capable of signalling the needs of both consumers and industrial users of agricultural outputs to the farmers. The marketing system must also motivate and reward all of the parties whose participation is essential to the delivery of commodities and products in the quantities and at the qualities demanded. Yet another development which has increased interest in marketing practices of late is the move towards market liberalisation as part of economic structural adjustment in many developing countries.

The marketing concept suggests that an organisation is best able to achieve its long term objectives by orientating all of its operations towards the task of consistently delivering satisfaction to the customer. In order to do so, the organisation must begin by getting to know what it is that will satisfy the customer. The marketing system as a whole has to be customer orientated. A marketing system comprises the functions of marketing (buying and selling, storage, transport and processing, and, standardisation of weights and measures, financing, risk bearing and market intelligence), and the organisations that perform them. Marketing systems have at least four sub-systems, these being production, distribution, consumption and regulation. These sub-systems often have conflicting interests that have to be resolved if the system as a whole is to be efficient and effective.

The food industry is a major user of agricultural products and commodities. As disposable incomes increase in developing countries, the food industry will have to meet new and different needs from its more affluent consumers. The food industry will, in turn, require agriculture support its efforts to meet the new challenges and opportunities. In particular, the food industry will demand that agriculture produces a wider range of qualities in its products and commodities with a greater proportion of total supply in the top grades; downward pressure will be exerted on agricultural production costs; agriculture will be required to supply throughout the year rather than seasonally; reliability in the quantity, quality and timing of supplies will become the major determinant in supplier selection; innovative producers who can provide differentiated products and products that make food processing easier or cheaper are more likely to survive than those who persist in producing traditional products using traditional farming methods; and issues related to the health aspects of food consumption will become increasingly important.

7

Agricultural Marketing System

Agriculture continues to be main stay of life for majority of the Indian population. It contributes around 25 per cent of the GDP and employs 65 per cent of the workforce in the country. Significant strides have been made in agriculture production during the last 50 years of independence. The agriculture production of food grains increased from 51 million tones in 1950-51 *i.e.* before beginning of the 1st Five Year Plan to 209 million tones in 1999-00. The output of oilseeds went up to 22 million tones. Similarly, the production of fruit and vegetables also increased to more than 134 million tones owing to the production efforts through all these years. The subject of agriculture and agricultural marketing is dealt with both by the States as well as the Central government in the country. Balanced regional development has been essential component of the development strategy and different Five Year Plans aimed at achieving socio-economic development of the country.

Starting from 1951, the different Five Year Plans laid stress on development of physical markets, on farm and off farm storage structures, facilities for standardisation and grading, packaging, transportation etc.. Development of horticulture marketing attracted attention of policy makers during the 3rd Five Year Plan. The year 1965 witnessed coming into existence of Central Warehousing Corporation, Food Corporation of India, Agricultural Prices Commission (later renamed as Commission for Agricultural Costs and Prices) and several other organisations.

Besides number of organisations under the aegis of Ministry of Agriculture, Consumer Affairs and Commerce were set up in the form of commodity boards, cooperative federations and export promotion councils for monitoring and boosting the production, consumption, marketing and export of various agricultural commodities. The prominent among them included Cotton Corporation of India Limited (CCI), the Jute Corporation of India Ltd. (JCI), the National Cooperative Development Corporation Ltd. (NCDC), the National Agricultural Cooperative Marketing Federation Ltd. (NAFED), the National Tobacco Growers Federation Ltd. (NTGF), the Tribal Cooperative Marketing Development Federation Ltd. (TRIFED), the National Consumers Cooperative Federation Ltd. (NCCF), etc for procurement and distribution of commodities; and the Tea Board, Coffee

Board, Coir Board, Rubber Board, Tobacco Board, Spices Board, Coconut Board, Central Silk Board, the National Dairy Development Board (NDDB), National Horticulture Board (NHB), State Trading Corporation (STC), Agricultural & Processed Foods Export Development Authority (APEDA), Marine Products Export Development Authority (MPEDA), the Indian Silk Export Promotion Council, the Cashewnuts Export Promotion Council of India (CEPC), etc. for promotion of production and exports of specific commodities.

Free Market Forces

Most agricultural commodity markets generally operate under the normal forces of demand and supply. However, with a view to protecting farmers' interest and to encourage them to increase production, the Government also fixes minimum support/statutory prices for some crops and makes arrangements for their purchase on state account whenever their price falls below the support level.

The role of Government normally is limited to protecting the interests of producers and consumers, only in respect of wage goods, mass consumption goods and essential goods. It is promoting organised marketing of agricultural commodities in the country through a network of regulated markets. To achieve an efficient system of buying and selling of agricultural commodities, most of the state Governments and Union Territories have enacted legislations to provide for development of agricultural produce markets. The basic objective of setting up of network of physical markets has been to ensure reasonable gain to the farmers by creating environment in markets for fair play of supply and demand forces, regulate market practices and attain transparency in transactions.

With a view to coping up with the need to handle increasing agricultural production, the number of regulated markets have also been increasing in the country. While by the end of 1950, there were 286 regulated markets in the country. Today the number stands at 7161(31.3.2001).

The Central Government advised all the State Governments to enact Marketing Legislation to provide competitive and transparent transactional methods to protect the interests of the farmers. Barring a few, most of the States and Union Territories embarked upon a massive programme of regulation of markets after enacting the legislation. Most of these regulated markets are wholesale markets. There are in all 7293 wholesale markets in the country. Besides, the country has 27294 rural periodical markets, about 15 per cent of which function under the ambit of regulation. The advent of regulated markets has helped in mitigating the market handicaps of producers/sellers at the wholesale assembling level. But, the rural periodic markets in general, and the tribal markets in particular, remained out of its developmental ambit.

Table. India: Projections of Production and Marketed Surplus of Agricultural Products (2006-07) (Million Tonnes)

Commodity	Marketed Surplus Ratio (%)	Production 1999-2000 (Estimated)	2006-07 (Projected)	Marketed Surplus 1999-2000 (Estimated)	2006-07 (Projected)
Rice	43.0	87.5	103.5	37.63	44.50
Wheat	51.5	68.7	84.3	35.38	43.41
Coarse Cereals	43.1	39.2	34.4	16.89	14.83
Total Cereals	-	195.4	222.2	89.90	102.74
Pulses	72.4	13.5	21.0	9.77	15.20
Total Food grains	-	208.9	243.2	99.67	117.94
Oilseeds	79.6	21.6	33.8	17.19	26.90
Groundnut	68.3	5.9	11.5	4.03	7.85
Mustard & Rape	84.3	6.1	8.9	5.14	7.50
Other oilseeds	86.3	9.6	13.4	8.28	11.56
Sugarcane	92.9	315.1	352.8	292.72	327.75
Cotton	100.0	2.1	3.2	2.10	3.20
Vegetables	83.0	85.0	110.7	70.55	91.88
Fruits	97.0	49.5	70.5	48.02	68.38

- *Market Surplus Ratio:* Sub-Group on Estimation of Marketed Surplus Ratio, Constituted by GOI.
- *Production Projection:* Kumar P. and V.C.Mathur, "Agriculture in *Future:* Demand - Supply Perspective for the Ninth Five Year Plan", Economic and Political Weekly, Sept., 28, 1996.

In India, agricultural marketing has more than 100 years' history of Government intervention, which goes back to British time. However, the objectives and forms of intervention have undergone a substantial change over time. It was envisaged that physical markets with facilities and services attract the farmers and the buyers creating competitive trade environment thereby offering best of the prices to the producers/sellers. While instances are common where the agricultural produce is marketed bypassing the regulated market yards, nevertheless, the physical markets as cardinal points of distribution will continue to remain important.

APMC and Boards

Agricultural Produce Marketing Committees (APMC) are corporate bodies established under the respective State Agricultural Produce Marketing Regulations Acts. They are either elected or nominated by the

Government from amongst representatives of agriculturists, traders and other functionaries and local representatives. All the State Acts provide for constitution of separate market committees for individual market except in Tamil Nadu where it is constituted at the district level to administer all the regulated markets in the district. The market committees are either controlled by the Director of Agricultural Marketing or the State Agricultural Marketing Board. States like Delhi, Uttar Pradesh, Rajasthan, Karnataka, Maharashtra, Gujarat, Madhya Pradesh, West Bengal and Orissa etc. have separate Marketing Department.

The Expert Committee observed that both the State Agricultural Marketing Departments as well as the Marketing Boards are functionally strong in U.P., M.P. and Rajasthan. In these States, regulatory functions have been entrusted to the State Marketing Departments while developmental functions are performed by the State Agricultural Marketing Boards. In Andhra Pradesh, Tamil Nadu and Karnataka, the Marketing Departments are functionally powerful, but the Boards are very weak and are more or less advisory in nature.

In Punjab, Haryana and Bihar, the Boards are powerful and attend to almost all functions and the Marketing Departments are either non-existent or weak. The Committee visualises that the State Agricultural Marketing Departments, Marketing Boards and Market Committees are like the vertices of a triangle which are vital and important for sustainable growth of the marketing sector. Their roles need to be clearly defined and demarcated to avoid any friction and impediments and to promote consistently progressive marketing management and development. The committee feels that a 'Code of Conduct' should be prepared for the market committees, as well as the boards and the State governments outlining their responsibilities, functioning and working conducive to efficient marketing system.

The State Agricultural Marketing Boards are represented by both official and non-official members. There is an imperative need to make the Boards administratively viable and managerially competent in keeping with liberalised trade atmosphere.

The marketing activities are many-fold and need liaison and collaboration with related organisations such as Railway Board, Forward Markets Commissions, Department of Posts & Telegraphs, Doordarshan, All India Radio, State Planning Commission/Board, consumers and farmers organisations, ports etc. It is desirable that the State Agricultural Marketing Board is given appropriate official recognition in these organisations so as to facilitate presentation of marketing activities and programmes of the State in their meetings and accelerate the pace of implementation of these programmes.

Although, market committees are, by and large, constituted democratically, they lack in professional management. The marketing manpower is controlled by three agencies, namely, Market Committee, State Marketing Department and State Marketing Board. Market Secretaries are yet to become market managers. The Committee strongly feels that all the market committees including sub-yards should be headed by professionals. Existing Secretaries need to be trained in professional management of the markets to facilitate liberalised, competitive and free marketing system.

Restrictive Regulated Markets

The institution of regulated markets, set up to strengthen and develop agricultural marketing in the country has achieved a limited success in providing transparent transactional methods/marketing practices, need based amenities and services conducive to efficient marketing. The restrictive legal provisions such as "all agricultural produce brought into or processed within market area shall pass through the principal market yard or sub market yard and shall not be bought or sold at any other place within the market area2" or "no such person shall carry on business as trader in agriculture produce into market area except in accordance with the license issued in this behalf by the Market committee3" did not augur well with competitive market structure. The power of the market committees to suspend and cancel the license granted to traders has become an impediment to the free structure.

The licensed traders also have not favored new entrants in the arena to maintain their grip over trade in market yard. The Agricultural Produce Market Committees have often yielded to unethical practices, which they were supposed to control and have instead become monopoly trading grounds. Promoting competition in trade and facilitating farmers with supporting services like grading, standardisation, storage with pledge finance and facilities in the market yards have become secondary activities. Even basic function of regulation, proper method of sale, correct weighment and prompt payment have eluded the market authorities.

The programme of development of markets in the country has remained confined mostly at the level of wholesale assembling markets. The regulated markets in the country generate income from market fee, license fee, rentals and miscellaneous income. The rate of market fee varies from 0.5 per cent to 2 per cent ad-valorum. The market Committees contribute 5 to 50 per cent of their income to the State Agricultural Marketing Board, depending on the relevant provisions, for development of market infrastructure in the State. The Rajasthan Government recently constituted a separate Market Development Fund, discontinuing the earlier practice of contribution to the Board by the Market Committees from their income. While these Boards undertook infrastructure development in respective States, instances are also common where funds from the Boards have been siphoned off to Public Ledger Account by the State Authorities.Consequently,modernisation/ infrastructure development conducive to operational efficiency of the markets has suffered heavily.

The state Governments are empowered to initiate the process of setting up of a market for certain commodities, which are regulated and for certain areas, in which the Regulation is enforced. As a result of this very process of initiation of a market, the service providers for agricultural marketing do not have any role. Nobody can take initiatives in assessing the viability and feasibility for setting up the markets equipped with the best facilities at competitive cost. Therefore, these provisions will have to be replaced by providing an omnibus provision that anybody can set up a market, provided minimum standards, specifications, formalities and procedures are complied with.

The purchaser should not be prevented from buying directly from the farmers. Products which are not locally grown and which have no substantial marketed surplus need not be included in notified commodities. Purchasers as well as sellers should be free to transact anywhere without attracting any of the provisions of APMC rules and byelaws. The Government of Karnataka has set the ball rolling by amending its Act to allow the National Dairy Development Board to set up wholesale fruit and vegetables market at Bangalore with the subsidiary collection centers etc. to promote integrated marketing of horticultural produce grown in the State.

The States, by an amendment, now should allow joint ventures or any other organisations, to set up markets in the interest of farmers, traders & consumers.

Although, technically the farmer/seller is free to sell his produce in any mandi he likes, practically he has no liberty to sell his produce in his village or to the retail chain/processor/bulk buyer directly. He has to take his produce to regulated market where the sales and deliveries are effected. This has hampered development of retail supply chain and direct supply to the processing, consuming factories or other bulk purchasers. Wherever there is a principal market, in the city or in the metropolitan, the sub-market or the collection center is not permitted, though in recent times "Rythu Bazar" or "Apni Mandis" or "Farmers' Mandis" are coming up on the peripheries of the metropolis, cities and towns. Therefore, the very basic concept of a single market, as a "principal market" for one agglomeration, has created monopoly in many respects, resulting a huge power center with plenty of exploitation practices.

The experience of the farmers, the consumers and the trade functionaries indicates that, under the existing legal provisions, there is no scope for:

- Direct marketing by the farmers or their groups to the retailers;
- The retailers can not directly approach the farmers or their groups in their farms or villages;
- The processors find it difficult to procure material at the production source and hence have to go the Mandis for procurement of the material;
- Electronic trading as the marketing includes trade by "electronic media", which is subject to the Regulation. Therefore, the electronic media traders have been shifting to the places, where there is no regulation, or where the regulation is not enforced. These markets resultantly have not seen any linkage with "futures" or "commodity exchanges" so far.

Licensing: Functionaries of the market yard such as commission agent, trader, processor, weighman, surveyor, broker, hamal, warehouse owner, transporters etc. to function in the area, as well as in the market or sub-market, are required to obtain a license. Instead of license, registration with APMC should be enough for any body to trade/operate in the market. Licensed functionaries in the markets such as traders, weighman, mathadis (Hamals), procurers, surveyor etc. have over a period acquired a monopoly status. New entrants are normally not permitted at the behest of respective associations and unions. The monopolies in marketing and handling have added to marketing costs detrimental to both producers and consumers. The typical example is that of transport charge from the principal market (Gultekdi Market yard) in Pune to various centers in the city. The transporters in principal market yard, Pune have unwritten agreement that no outsider would be allowed to enter into transport activity. As a result, transport charges from Pune Principal Market for a distance of about 3 or 5 kms. are more than transport from upcountry areas like Solapur, Nasik, etc. In the process, farmer suffers most.

The Rules and Bye-laws stipulate lot of procedures and documentation for licensing. The documentation mainly consists of Solvency Certificate, Certificate of good behaviour, cash security, Bank Guarantee or third party guarantee, etc. The License is granted for the area, but they have to operate in the market or the sub-market.

Moreover, the license is granted for only one year, which has to be renewed. Renewal takes time and business is hampered. Therefore, only registration with APMC should be there as has been suggested earlier. As the Rule provides that no one shall market any declared agricultural produce in the market area, other than the principal yard or sub-yard. The very fact of licensing has converted the entire scenario into handing over the monopoly to those, who have already got the license. The new-comers are normally kept aside on the ground that there is no potential of additional business. This monopolistic circumstance has caused lot of hardships to the producers and the consumers. They disallow latest systems of handling, like cleaning, grading, packaging, weighing, transporting, etc. Their organisations, associations and unions have become so strong that they can dictate their terms to anybody, including the Government. Thus, the element of cost efficiency and competitiveness is lost because of licensing.

Neglected Rural Haats

The present haats or weekly bazaars which constitute first contact point with commercial circuits for the producers have not been provided with amenities and facilities required for operational, technical and pricing efficiency. Improvement in the efficiency of the rural markets has direct impact on farmers' level of income. These markets need improved services for users to facilitate marketing of the local produce, creating an element of market security for the growers. The rural market can also be used for effective credit, input marketing and procurement activities. Haats need to be provided with mobile banks on haat days by Gramin Banks.

Even though the 73rd Amendment to Constitution bestowed management of rural markets to village panchayats, different authorities still continue to own and operate these markets. Moreover, where these markets are brought under regulation as sub yards, and have turned to a case of multiple collection of ground rent, market fee and other charges etc. by different agencies disregarding their developmental responsibilities. The fees or other charges collected in these markets are also disproportionate to the amenities and facilities rendered.

The rural periodic markets, besides transacting agricultural and allied commodities, also perform the function of distribution of items of daily needs of the rural population. Farmer is a seller, consumer and even trader in these grass root level markets. 80 per cent of the household income of the rural masses is estimated to be spent at these markets. Their development, therefore, with proper operational, pricing and technical efficiency constitute foundation of integrated market system for distribution of agricultural and allied produce. Owing to their financial non-viability, these markets did not receive adequate attention either from the State or Central Government for their developmental requirements.

Their development should be looked as socio-economic development as multipurpose growth centers so that these places could be used for other developmental activities related to health, education, animal husbandry on days other than haat day and a scheme to this effect in the sector is, therefore, necessary.

Multiple Legal Instruments

Apart from the Agricultural Produce Markets Acts, activities of market functionaries are regulated by several other legal instruments promulgated by the Central government and the States.

The important ones are:

- The Indian Sale of Goods Act 1930.
- Agricultural Produce (Grading and Marking) Act, 1937, 1986
- Drugs and Cosmetics Act 1940.
- Vegetable Oil Products (Control) Order, 1947.
- The Emblems and Names (Prevention of Improper use) Act 1950.
- Prevention of Food Adulteration (PFA) Act, 1954, 1964, 1976, 1986
- The Essential Commodities Act, 1955
- The Sugar Control Order 1956.
- The Export (Quality Control & Inspection) Act 1963.
- Fruit Products Order (FPO), 1965, 1997.
- Solvent Extracted Oil, Deoiled Meal and Edible Oil (control) Order, 1967
- Monopolies and Restrictive Trade Act, 1969.
- Meat Food Products Order, 1973
- Vegetable Oil Products (Standards of Quality) Order, 1975.
- The Packaged Commodities Order 1975.
- Standards of Weights and Measures Act, 1976
- Pulses, Edible Oilseeds and Edible Oils (storage control) Order, 1977
- Consumer Protection Act, 1986
- Cotton (Control) Order, 1986
- Milk and Milk Products Order, 1992
- Bureau of Indian Standards Act 1986.
- Jute and Jute Textiles (Control) Order, 2000.
- The Trade and Merchandise Marks Act.
- The Town & Country Planning Act.
- The Muncipalities Act.
- Labour Laws.

These apart, in exercise of the inherent powers of the Central Government and delegated powers to the States/Uts, various orders in respect of foodgrains have been made by the Central/State Governments relating to: (i) licensing of dealers/retailers for trade in foodgrains; (ii) regulation of stock limits; (iii) restrictions on movement of foodgrains; (iv) compulsory purchase by the Government under the system of levy. The recent position with regard to restrictions on movement of food and agricultural produce reveals that some of the States have impsed restrictions on movement of some essential commodities.

Most of these restrictions relate to foodgrains and more particularly to rice/paddy. States like Andhra Pradesh, Tamil Nadu, Karnataka, J&K and Madhya Pradesh have imposed statutory restrictions on movement of rice/paddy outside the State with a view to maximizing procurement and to meet internal requirements (though Government of Andhra Pradesh has kept in abeyance the order imposing restrictions on movement of paddy from 27.7.2000). Besides the statutory restrictions, some States also imposed informal restrictions on movement of foodgrains outside the state during particular periods of the year.

In the light of the present position with regard to production and supply of foodgrains in the country, it is essential to review all such restrictions. With the removal of restrictions on stocking and storage, warehousing activities will get a fillip and stored foodgrains acting as a collateral will facilitate marketing credit or pledge finance for producers, as well as entrepreneurs. In the light of 'one world one market' there is no use keeping such restrictions. In order that market should be free and competitive, all restrictions must go. The provisions under the legal instruments, are used particularly during the periods of short supply as reflected through high prices to regulate the activities of traders and processors pertaining to trading, stocking, maintenance of quality, grading, packing, processing, blending and movements.

These instruments are administered by different Ministries and Departments of the Central government and the States. Considering the fact that shortages have now been replaced by surpluses, these restrictive provisions have to be put to an end for uninhibited free marketing system. The Essential Commodities Act, 1955 needs to be repealed in the new context.

Linking Spot Markets with Futures Markets

The regulated markets, which are spot markets, have not been able to link themselves with the forward and future markets to receive price signals. The futures trading in agricultural commodities is also regulated by the Government. Till recently, it was permitted in jaggery, blackpepper, turmeric and hessian. The Kabra Committee appointed by the Government of India in 1993 recommended for permitting futures trading in 17 major agricultural commodities. The Government has now notified the introduction of futures trading in cotton, kapas, raw jute and jute goods, all major oilseeds and their oils and cakes; rice bran oil and coffee. The recent addition to this list is sugar. The Committee is of the view that more and more commodities should be added to the list of commodities allowed for futures trading to facilitate integration of domestic market with international market.

Domestic Market Reforms

Reforms in the domestic markets for farm products have to precede trade liberalisation. Certain provisions, which come in the way of efficient functioning of the domestic market for agricultural commodities and adversely affect both the growers and the consumers, should be done away with in phases. These are levy on rice millers, statutory rationing of rice and wheat inCalcutta, monopsony procurement of raw cotton in Maharashtra, levy on sugar mills, system of state advised prices of sugarcane prevalent in some states, stocking limits particularly in post-harvest season etc.

The food grains marketing system has been a subject of considerable debate in recent years. In this connection, it needs to be noted that the intervention by the government in food grains markets, specially of rice and wheat, in the form of price support, buffer stocking and P.D.S., has helped the country in achieving near self-sufficiency in staple food and in improving physical and economic access of masses to food. The Committee recommends that the food security and food management system built up over the years should be retained as an essential component of national security policy. For this purpose, a multi agency approach along with Food Corporation of India may be adopted in the national food management system.

Government Marketing Support Programme

Because of heavy dependence on nature, the production of agricultural commodities is uncertain. Sometimes good harvest causes scenario of plenty's, while the poor harvest creates scarcity or short supplies. Both these situations adversely affect the farmers. Government's minimum price support scheme in this context has remained very useful to the farmers. The minimum price support policy is applicable to 24 agricultural commodities. Under this scheme government agencies are asked to start purchase of the commodities when market prices fall below the minimum support price so as to save farmers from the distress sales.

Though scheme has proved very useful to the farmers, its implementation has remained faulty and inefficient. The agencies involved in such purchases do not start operations in time. They do not make payments in time. They do not have sufficient arrangement for weighing, bagging, storage, documentation and payment of purchases. Sales of purchased commodities is normally not done in time and with marketing efficiencies. Therefore heavy losses do occur. However it is generally viewed that government intervention in food grains specially rice and wheat, in this form of price support, buffer stocking and PDS has helped the country in achieving self sufficiency in staple food and in improving economic access of masses to food while protection the interests of the producers. The committee therefore recommends the continuation of the food security and food management system built up over the years as essential component of National Food Security. However the agencies involved in this management should be encouraged for optimum efficiency and diligence.

Market Intervention is another marketing support policy of the government. Over and above the commodities under minimum support price, the prices of some commodities especially of horticultural crops tend to fall drastically forcing the farmers for distress sales. Under these circumstances with the help of state government, GOI launches Market Intervention Scheme for that particular crop in that season so as to avoid distress sales by the farmers. Prices based on the cost of production and other factors for that season are decided for Market Intervention. Government agencies are assign the job of intervention. The losses are shared equally by Government of India and State Government.

This policy of Market Intervention also has proved a boon to the farmers in distress. The operational efficiency of purchasing agencies need to be toned up in the context of cost efficient purchases vis -a -vis competitive sales so as to avoid or reduce losses. Therefore the Committee not only recommends continuation of Market Intervention Scheme but also suggests expanding the coverage of this scheme to more commodities. The Expert Committee also recommends that the Government of India may encourage the state government to initiated market intervention operations well in advance for saving the farmers in distress.

The experience shows that while in most of the surplus producing regions, the state governments and their agencies remain active for effective implementation of the policy of assuring minimum support prices, in those areas where the need for price support arises only once in two or three years, the public agencies have not been able to provide effective support to the farmers as they could not tie up with central nodal agency for making necessary purchase arrangements in time. Such failures on the part of the state agencies lead to a set back to the production programmes. Most of the upcoming regions like Eastern Uttar Pradesh, Madhya Pradesh, West Bengal, Orissa, Bihar and parts of other states are likely to experience such situation more frequently. And these are the areas where there is considerable scope to increase the yield levels. It is in this context that for accelerating the production of food grains and other agricultural commodities, not only the market infrastructure needs to be strengthened but the price support policy also needs to be effectively implemented in all the regions of the country.

The minimum price support policy is currently applicable to 24 agricultural commodities which should be continued. However, there are several other commodities which are of considerable economic significance in various agro-climatic regions of the country. The fluctuations in their prices are considerably more. For protecting both the producers and consumers against wide fluctuations in the prices, there is a need for initiating a comprehensive price stabilisation scheme applicable to selected important crops of each agro-climatic region. The scope and contents of market invention scheme of Ministry of Agriculture be expanded and state governments should be encouraged to formulate and implement market intervention programmes for establishing the prices of crops not covered by MSP policy.

There is a considerable variation in the structure of taxes and fee on the agricultural produce in various states which also distorts the operation of the domestic market, gives wrong signals to the producers and leads to considerable fallacy in the efficiency of the operation of private trade vis-à-vis farmers cooperatives and public agencies. There is a need for bringing uniformity in the state level tax structure for agricultural commodities for improving the marketing efficiency.

India has created considerable infrastructure at the wholesale market level in terms of market yards with associated facilities, approach roads and market committees with representation of farmers. These committees are supervised by the state level marketing boards and state Departments of Agricultural Marketing. These organisations have helped in orderly marketing of the farmers produce. However, there is a need for redefining their role and structure in the context of emerging economic environment and market reforms. The emphasis now, should shift to other important activities of topical relevance. Some of the activities which should receive increasing attention of APMCs and departments of agricultural marketing include: (a) creating cleaning, grading and packing facilities; (b) encouraging the local farmers, traders and processors to market the produce under their brands and promoting such local brands based on graded produce; (c) creating minimum infrastructure in all, around 28 thousand rural periodic market places (haats) and linking these with other markets through roads; and (d) in the event of prices of any commodity, which is covered under the price support programme, tending to dip below the support level, either ensuring that an agency is in place to make purchases at the support price, or entering into the market on behalf of the nodal agency to prevent the prices from falling below the support level.

The objectives and forms of intervention in the marketing system have undergone a substantial change from mid sixties when the country opted for a package of interventions for protecting and reconciling the interests of consumers/producers as well as the industry. One of the important instruments used by the Government to intervene in produce markets consists of fixation and announcement of administered prices and arrangements for their implementation. The administered price regime currently in vogue includes (a) minimum support prices (MSP) for 23 commodities (7 cereals, 4 pulses, 8 oilseeds, copra, raw cotton, raw jute and VFC tobacco); (b) statutory minimum prices for sugarcane; (c) levy prices for rice; and (d) central issue prices for rice, wheat and coarse cereals for sale under public distribution system. Direct entry of public agencies in the marketing have influence on its structure, conduct and performance. Maintenance of stock of rice and wheat; distribution of cereals and sugar at prices lower than market prices; and open market operations by public agencies cast their influences on market.

Poor Credit Flow

Poor credit flow to agriculture and wholesale trade has been one of the major bottlenecks in the country's agricultural marketing system. Most of the lending to the agriculture sector is short term, and is in the form of crop loans. Farmers are expected to buy all the required inputs to raise a crop with the credit made available to them and then return all loans soon after harvest. Post-harvest credit from banks is usually not available. Given the importance of a good credit history, most farmers, especially small farmers, take the produce to market soon after harvest. Most farmers face the same objective function and produce is brought to market simultaneously by several farmers. As a result of the temporal supply conditions, prices fall. Often such fall in prices leave little surplus for the next crop.

The wholesale trade combines several functions. Financing is one of these functions rendered by the Commission agents and traders known as adatiyas in many parts of the country. Kuccha adatiyas are provided finance by pucca adatiyas to make payment to the farmers. Where produce can be carried in inventory with associated incentives, farmers often access credit from the wholesale trade. Pucca adatiyas often carry inventory consisting of produce sold by farmers. The informal sector provides significant credit to agriculture and wholesale trade, but the cost of credit is high compared to the rate at which it may be provided by banks. Lending against stocks or inventory of produce is a significant activity of banks. For example, banks lend to cotton spinning mills against stocks of cotton. However, bank credit to farmers against agriculture produce is quite uncommon. These lacunne should be corrected.

Post-Harvest Losses

Although, India produces a wide variety of fruits and vegetables, in the absence of adequate post harvest and marketing infrastructure *viz.* irradiation facilities, storage/cold storage and cold chain facilities, the horticulture produce suffer heavy post harvest losses. The problem is further complicated due to the fact that there are no storage facilities at the farm level and the farmers are forced to dispose off the entire produce immediately on harvesting. This creates a glut situation in the market. The margins of the wholesalers and retailers are therefore much higher in our country. The entire burden of the distress sale in the form of low prices falls on farmers.

The horticulture marketing practices lack systems approach. Trading and marketing structure is traditional consisting of a long chain of intermediaries. As many as 75 per cent of the farmers sell their produce at the farm level. They cannot afford to go to distant mandis on account of lack of facilities, expensive transportation and malpractices in the assembling markets. The long marketing channel is detrimental to quality and safety of those perishable products. It is essential to shorten the channels. As has been suggested earlier these commodities can be taken out of the exclusive purview of State Agricultural Marketing legislation to facilitate direct sales and other competitive outlet after proper grading, packing, transportation such as through refer vans to the final destination. The multiple handling add costs and increase post-harvest losses, adversely affecting the income of farmers and making produce comparatively costlier to the consumers. To minimize these losses, improvements are required at various levels, *viz.*: harvesting; grading; transport; storage; processing, packing and marketing.

Need for Legal Reforms

The nature of legal framework within which agricultural markets operate has a fundamental effect on the functioning of the agricultural marketing system. Legal reforms can play an important role in making the marketing system more effective and efficient by removing unnecessary restrictions and by establishing a sound framework to reduce uncertainty of the market. The volatility of commodity prices has been one of the main justifications for State intervention to control or direct the functioning of agricultural markets. Today most developing countries have adopted national policy committed to the liberalisation of their domestic agricultural marketing systems and to encourage an efficient and competitive private sector. In South Korea, for instance, a Task Force for Regulating Reform was set up and on the basis of its recommendations, 75.2 per cent of total of 701 regulating items in agricultural sector were either improved or abolished.

In promoting private sector marketing systems, Government need to examine existing policies, rules and regulations with a view to minimizing conflict in successful private sector operations. A review is required in respect of all laws which regulate participation in market such as registration/licensing, commodities traded, controls on packaging and labeling, laws affecting market place, laws affecting supply including controls on movement of produce and volume of commodities traded. The Expert Committee recommends that a Task Force be set up under the Ministry of Agriculture, Department of Agriculture & Cooperation to undertake a review of all marketing legislations and suggest introduction of necessary legal reforms to promote free and fair marketing system for agricultural and allied products.

Strategic Approach for Agricultural Marketing

Agricultural Marketing

Marketing of the form produce begins with the planning of producing the commodity. Agricultural commodities besides being seasonal and regional in nature are bulky and the consumption of the same is spread throughout the year. Marketing is sum of all activities, which link the producer and consumer.

Strategic Approach for Marketing

Marketing of farm produce needs a strategy, in order to establish a bridge between producer and consumer efficiently and economically. Marketing strategy comprises defined specific objective and action plan to have competitive edge over other players in this area. Prudent producer will analyse the strategy thoroughly, to take up farming to maximize his out put in respect of quantity and quality with cost effective approach for a competitive marketing to drive a reasonable return over cost invested.

Since the last two decades, India has not only attained self- sufficiency in food grain requirement, but has become an important player in the export market. Keeping in view the domestic requirement, our farmers have been guided very often for producing meticulously market-oriented production. Market oriented production will have to be taken up to match the demand and supply equitably.

Marketing Environment

Soon after harvesting, farmer intends to shift the produce to market, as early as possible. Farmer expects an environment to market the produce, where his decision should not be effected by any other variable factor or forces and is assured of a competitive price. This objective is enshrined in the creation of regulated market. In the management of these markets, a farmer representative serves as a member and this gives him a sense of belonging in the market, which makes the market environment more conducive for completing the transactions.

Marketing environment also includes, required infrastructure for efficient handling of the commodity, such as covered auction platform, tested and trusted weighing machines, cleaning, grading and storage/ cold storage facility, loading and unloading facilities, internal roads to decongest the vehicular movement, pledge finance availability, banking, postal services, cattle-shade, farmer's rest house and link road from farm to the market. Trained manpower in these markets increases the conduciveness of the marketing environment. Besides these requirements, big institutional support such as NAFED, CWC/SWC, DMI, NHB, Directorate of Economics and Statistics, NABARD, APEDA, Commodity Boards, CACP, Forward Market Commission and State Agricultural Marketing Boards - provide much needed fillip to the marketing environment.

Alternative Marketing

The traditional marketing system of farm produce has paved way for other alternative marketing system, which is more need based and sometimes suits the individual partners in the trade. The need of the alternative marketing of farm produce is to reduce the distributive system. Another reason is to explore the possibility of bringing the buyer to the production place, that is, bringing the market at farm.

Some of such marketing systems are briefly given below:

- *Group Marketing:* Group marketing includes, joint planning, funding, implementation, pricing, sharing risk equally in the marketing and reaping the benefit of collective bargaining. This indirectly results into more returns and economises the marketing cost.
- *Cooperative Marketing:* This marketing system is pursued through cooperative societies registered under Cooperative Societies Act. This system is pursued on the principle of "self help by mutual

help". It reduces the marketing cost, enhances the bargaining power and there is equitable distribution of the proceeds.presence of marketing cooperatives makes the market more competitive and ensures better returns to the producer. This system is owned and managed by the farmer themselves for their own economic betterment and enhancing marketing efficiency. At the village level, a large network of multipurpose/ commodity specific Primary Agricultural cooperative Societies are supporting such marketing system in the country.

- *Direct Marketing:* Farmers come into direct contact with the consumers and receive the payment directly from the consumers. *This system is prevailing in many parts of the country viz:*
  - *Apni Mandi:* In these mandies, farmer-producers bring the produce for sale directly to the consumers,as in Punjab and Rajasthan
  - *Rythu Bazrs:* In Andhra Pradesh, such markets have been established to provide direct link between farmers and consumers in the marketing of fruit and vegetables and other essential food items.
  - *Uzhavar Santhaigal:* Government of Tamil Nadu had established farmers markets called Uzhavar Sandies in selected municipal and panchayat areas.In these markets, better marketing infrastructure is provided to farmers free of cost. Farmer get other inputs and quality seeds in these markets.(However it has been discontinued by the present State Government)
  - *Raithara Santhegalu:* In Karnataka, marketing Board has established farmes market or Raithara Santhe without any middle men and provide a place where farmers and consumers directly interact.
- *Contract Farming:* Contract farming, is a type of farming wherein the industry or perspective buyer enters into a contract with the farmer and promises to buy the farmer's produce at a pre-negotiated price under pre-negotiated conditions. Besides, the buyer agrees to supply the required farm inputs at the required time. In this process farmers are assured of an established market and a fixed price for their produce. The buyers would be able to procure the produce of a specified quality at much cheaper rate.

Normally contract farming involves the following basic elements:

- Pre-agreed price
- Quality

- Quantity or acreage (minimum/ maximum)
- Time
  - Farmer is contracted to plant the contractor's crop on his land.
  - Harvest and deliver to the contractor, a quantum of produce, based upon anticipated yield and contracted acreage.
  - This could be at pre-agreed price.
  - Towards these ends, the contractor can supply the farmer with selected inputs

Thus, under the contract farming, contractor supplies all the inputs required for cultivation, while the farmer supplies land and labour.

ADVANTAGES OF CONTRACT FARMING

To the farmers:

- Assured market and support price.
- Efficient timely technical guidance- almost free of cost. Crop monitoring on a regular basis.
- Financial support in kind.
- Assured quality of seeds and pesticides.
- Better price for produce- No Middlemen.
- Gain of bulk supply instead of small lots.
- Remunerative returns and timely payment.

To the buyers:

- Backward market integration is possible with assured supply.
- Ensured required quality- ensures residual toxicity level.
- Uninterrupted and regular flow of quality raw material
- Protections from fluctuation in market pricing.
- Buyers can plan on long term basis.
- Buyers select the products, which have international demand.
- Selection of location- agroclimatic conditions matching with the commodity's requirement and less prone to natural calamities.

Success Of Cotract Farming Lies In:

- Buyer's ability to implement the plan.
- Monitoring the growth of the plants package of practices, pest management and sanitation measures
- On site procurement- processing, packing and forwarding.

Indian Experience in Contract farming:

- In Punjab and Rajasthan-Tomato Pulp-Pepsi

- In Maharashtra and AP-Exotic Vegetables-Trikaya foods/VST and small farmers.
- In Bangalore-Gherkins-Exporter with farmer growers
- In AP/Karnataka-Edible oil and Sunflower-ITC Agro -tech

Agri-Clinics and Agri-Business centers

Providing testing facilities, diagnostic and control services and other consultancies on a fee for service basis though nearly 5000 such clinics.The programme is implemented jintly through Small Farmers Agribusiness Consortium (SAFC) and National Institute of Agricultural Extension management (MANAGE). These services are provided by well trained graduates and financial support on subsidised manner is being provided by NABARD to establish the needed infrastructure.

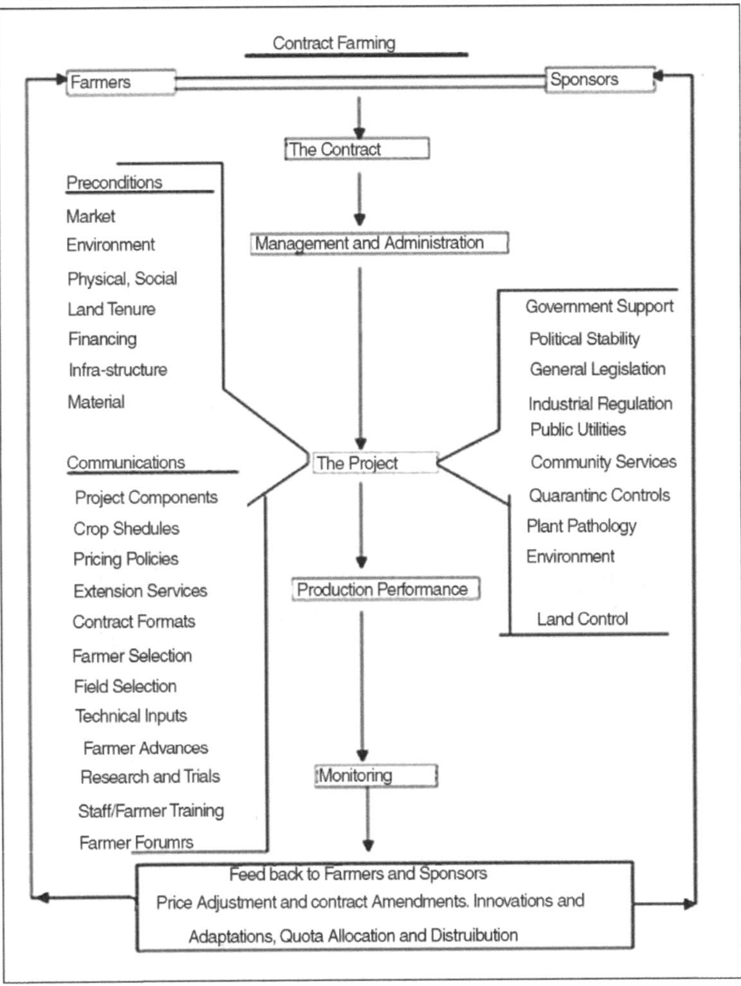

Futures Trading/Forward Contracts:

- Bombay Cotton Traders Assocation (1875), for regulation of cotton trade. Futures Markets were established for oil seeds in Bombay (1900) for Wheat in Hapur (1913) raw jute and jute goods in Calcutta (1912)
- Forward Contract(Regulation) ACT - 1952 - to check the unhealthy Speculation - forward trading was regulated and prohibited in other areas.
- Forward Market Commission -Established in 1953. The Spices and Oil seeds Exchange Ltd. Was established Sangli in-1953.
- Central Government has powers to notify the commodities and jurisdiction in

Which forward contracts are regulated.

- There is ban in about 100 commodities on forward trading.
- Two broad categories of Operators are namely., Hedgers and Speculators.

Risks in agricultural marketing:

- Qualitative and quantitative
- Institutional
- Price risk

Price Risk: Speculation and hedging to overcome this risk.

Speculation: absolutely profit motive

Hedging: executing opposite sales or purchases in futures market to offset the purchases or sales of physical product made in the cash market.

Future Trading: also known as Hedge Market.

K.N.Kabra Committee- 1993 - Reviewed the working of Forward Markets. Allowed futures trading in 17 commodities, for example, cotton, jute, custard oil, pepper and 11 oil seeds, oils and their cakes.

At present, in custard seed, turmeric, pepper, gur (jaggery), potato and hessian, future trading is undertaken. Now future trading is permitted in more than 100 commodities.

Types of Contract:

- *Ready contracts:* when delivery of goods and payment thereof, if completed 11 days from the date of contract. Not covered under the Act.
- *Forward contracts:* when delivery of goods that are not ready-delivery contract, and are governed by the Act.Two types-(1) Specific Delivery Contracts (2) Non- Contracts Specific delivery Contracts- are of two types on the basis of transferability of the rights or/

obligations, which is permitted in transferable specific contracts -TSD, and not permitted in Non- Transferable Specific Delivery Contracts-(NTSD)-(also called Futures Contract)

- *Option Contract:* for price insurance for producers These contracts are negotiated directly between the parties depending on availability, requirement and negotiated contract terms, such as, quality, price, period of delivery, place of delivery, payment terms etc., are incorporated in the contracts entered into These contracts are entered under the auspices of EXCHANGE OR ASSOCIATION.
- *Option in Goods:* agreement for purchase or sale of a right to buy or sell, or a right to buy and sell goods in future and include a Teji, a Mandi, a Teji Mandi, a Galli, a Put, a Call, or a Put and Call in goods.

Option in goods is totally prohibited under the Act. A Put - an option that gives the option buyers the right to sell particular futures contract at a specific price.

A Call - an option that gives the buyers the right to purchase a future contract at a specific price.

Characteristics of Future Trading:

- Futures Contact is highly standardised contract.
- Organised through exchange/association.
- Contract entered is for standard variety known as Basic Variety. With permission to deliver other identified varieties known as Tenderable Varieties.
- Units of price quotation and trading are fixed. Not alterable.
- Delivery period specified.
- Seller may deliver the gods at exchange or other pre-specified centres.
- In Futures Market, actual delivery of goods takes place only in a few cases.
- Mostly transactions are squared up before due date of maturity of contract and contracts are settled by payment of differences without any physical delivery of goods.

Advantages:

- Price discovery and price risk management with reference to the given commodity.
- Producer can get an idea of price likely to prevail at a future point.
- Consumer can also get an idea of the price at which commodity would be available.

- Useful to exporters, as they get advance indication of the price likely to prevail and help the exporter in quoting the realistic price. Hedging of risk in futures market.
- Price stabilisation, leads to integrated price structure.
- Facilitates production and manufacturing activities.
- Maintains balance in supply and demand.
- Encourages competition and acts as price barometer.

Charcteristics of commodities for futures trading:

- Plentiful supply.
- Must be storable - least perishable to undertake future delivery.
- Homogeneous, gradable - future delivery without dispute in quality.
- Large demand from large number of consumers.
- A few large firms should not control supply of commodity.

On Line Commodity Exchange Of India Limited (Oceil).In Ahmedabad, started functioning on 26th Nov 2002, India's first demutualised, on line multi- commodity exchange.Electronic trading and demutualised system makes the exchange uique,ensures flawless trading, enhancing the confidence of the trade participants. In demutualised setup, the ownership, management and trading are in the hands of three different sets of people. This completely eliminates any conflict of interest and helps in pursuing policies and practices.

Salient features of exchange:

- Convergence of all offers and bids emanating from all over the country in a single electronic order book of the exchange.
- Participation of importers, exporters, grower, brokers, traders etc. using an electronic trading system.
- Fair trading practice through checks and balances built in the system.
- Trading with the help of information technology and communication network.
- Efficient guaranteed clearing and settlement system enabling book entry settlement.
- Warehouse receipt system based delivery of underlying commodities.
- Real time price trade data dissemination
- Reliable, effective and impartial rule based management by professionals having no trade interest.
- On-line position monitoring for contracts. Margin calculation,

monitoring of positions of the trading and clearing members. OCEIL has an in-house ClearingHouse at AHMEDABAD. It has connectivity with all members and the clearing banks There is a provision of adequate margin to ensure that the chances of default are reduced.The clearing house is the counter party to each trade and will be responsible to and reporting to the Exchange Officials. A computerised system ensures the on-line calculation of margin. The position is marked to the end of the every trading session to calculate the profit or loss at any given point of time. The clearing bank, having a nationwide network is equipped with electronic fund transfer facility. The Exchange has a trade guarantee fund and proposes to setup a customer protection fund.

Future Planning:

- To become National Commodity Exchange.
- To emerge as the largest commodity exchange in the country.
- Consolidation of existing commodity exchanges.
- Trading in more commodities including sugar.
- Greater involvement of institutions.
- Reaching every corner of the country

OIECL-s strategy is to focus on the securitisation of commodities through warehouse receipts system, electronic transfer of stocks in demat form, interface of on-line trading of such stocks to be linked with interest rates and to focus on the participation of commodity funds for investment and banks and institutions for both, finance & investment.

Pledge Finance Scheme

On the recommendations made by the All India Rural Credit Survey Report of Reserve Bank of India, in 1954, pledge concept was developed to help the producers to come out of the clutches of the village money lenders and to avoid distress sale and situation of glut in the market.In lieu of the produce stored with the different agencies, finance was made available to the extent of 70 to 80 per cent of the value of the produce on a very low rate of interest.Pledge loan facility is being provided by CWC/ SWC, Agricultural Produce Market Committees in many states and Rural godown Scheme or Gramin Bhandaran Yojana of Directorate of Marketing and Inspection, govt. of India.

Recently, NAFED has also launched anew scheme to extend pledge loan facility to small and marginal farmers against the stock stored in NAFED's societies' godown. Under this scheme, small and marginal farmers will be immediately advanced an amount of up to 80 per cent of the assessed value of the stock on the given day for non-perishable stock and up to 60 per cent for perishable stock, against hypothecation or pledge of stock to NAFED.An interest rate of one per cent per month will be charged by NAFED for this service.

Under this scheme the produce must be of defined quality or graded before taken into possession in the warehouses. Quality of the stored produce is maintained on scientific lines to prevent the storage loses and delivers the same to the depositor on his request.

Legal Reforms

As in the other sectors of the economy, process of reforms in agricultural marketing has started with the liberalisation. Its initiation was first observed with withdrawal of compulsory quality control on notified commodities by Govt.of India.To maintain and ascertain the quality of the produce for export was left with the exporter trader.

The ban lifted over the interstate movement of the agricultural produce has provided an opportunity to the farmers to dispose of their produce more profitably anywhere in the country.

To avoid the risk due to price fluctuations, more and more agricultural commodities have been notified by the government of India and brought under the ambit of Forward Market Commission for forward/ futures trading. Directorate of Marketing and Inspection, Govt.of India, has prepared a model State Agricultural Produce market Act, and proposed certain amendments to provide privatisation of Agri-Business, bringing uniformity in the marketing environment throughout the country and to provide better infrastructure to handle the farm produce and promote contract farming with legal protection.

For the international trade in agricultural commodities and agr-products, quantitative restrictions have been removed in many commodities and they have been brought under the Open General License (OGL) category. Such reforms in the age of open economy are imperative for the promotion of farm produce in the international market and grab the opportunity for getting more profits by entering the new markets.

E. COMMERCE

According to WTO E. Commerce is production, distribution, marketing, sale or delivery of goods and services by electronic means. The world is becoming one whole market.The E.Com. environment is going to be charecterised by free flow of trade.

Commercial transactions may be divided into three main stages:

- Advertising & searching stage
- Ordering & payment stage
- Delivery stage.

E Com is an advance state of electronic technology. It is just an extension of business conducted on net. It is an interactive TV with Internet taking over computing sphere. A complete reinvention of how one does business through E.Com. makes communication, information gathering & trade between companies and consumer easier and faster.

E. Com can take two forms:

- Direct E. COM, Under this product or service such as music or professional legal advice, delivered to the buyer.
- Indirect E.Com: Here the product is ordered on the net but it is delivered in the normal way eg Book Supply com.

India got internet connectivity in 1989. There is a good scope for expanding customer base, coupled with quicker service & immediate delivery. Advantages of E COM-

- Sale of specialised products to affluent sections- retail sellers on E COM can sell specialised & high priced products that appeals to an audience of affluent society
- Wider access to customer globally at a low marginal cost.
- Sale to institutional members- who have embraced Web most conventionally.
- Closer relationship- buyers and sellers can build closer relationship electronically.
- Suitability to certain products eg. Soft ware, market research and sports travel can get immensely benefited from E COM.
- Suitability for the products affected by changes, as they offer only current products on the site-adjusting price in real time in response to fluctuations in demand.
- Global prices among productive units, zero transaction cost & no barriers to entry, E com. Economy comes quite close to these features of perfect competition. As large number of buyers and sellers can instantly interact with each other.
- Consumers can compare the prices of the products of different sellers and obtain goods at lowest quote.
- Development of Cybermediaries.When consumers make a purchase transaction, these cybermediaries record personal details of the consumers (eg. Consumers income group, their preference for brand, colour, size etc.) and process & analyse this information and new products are developed accordingly.As such, intermediaries are being replaced by the cybermediaries.

- Smaller production units can easily launch a web site and compete with large firms of the real world.
- Free flow of information is expected to result in efficient allocation of resources.
- Transaction cost is almost nil for products, which can be converted into digital form and can be supplied online.

Problems of E COM:

- Indispensability and connection with banking system-business must be linked.
- Growth of the credit card culture- the faster the growth of credit card, the faster would gain consumer acceptance and adoption.
- Absence of retail marketing- this is one of the precondition for adoption of E COM.
- Language problem.
- Scope for fraud- transactions being impersonal, anonymous and automatic can be manipulated easily.
- Absence of legal frame work- like the model law of E COM formulated by the UN Commission on international trade law in 1996. Govt. can effectively Tax E.Com transactions, but tax collection authorities do not have the powers to tax products, which are sold or produced beyond the local/ national boundaries. Since domestic consumers can access the global market, tax collection will be crucial.
- Backup service and updating web page.
- Unsuitability to certain industries *e.g.* foodgrain, cosmetics, perfumes

Rural Marketing: It's Potential, Importance, Problems and Distribution Strategy

Rural marketing involves addressing over 700 million potential consumers and over 40 per cent of the Indian middle income. No wonder, the rural markets have been a vital source of growth for most companies. For a number of PMCG companies is the country, more than half their annual sales come from the rural market.

Rural Marketing Potential in India

While we all accept that the heart or India lives in its villages and the Indian rural market with its vast size and demand base offers great opportunities to marketers, we tend to conclude that the purse does not stay with them. Rural marketing involves addressing over 700 million potential consumers and over 40 per cent of the Indian middle income. No wonder, the rural markets have been a vital source of growth for most companies. For a number of PMCG companies is the country, more than half their annual sales come from the rural market.

Among various media of communication, television and radio have played prominent rules in the rural India to-day. In the South, the penetration of satellite television is very high. Due to globalisation, economic liberalisation, IT revolution, female power, and improving infrastructure, middle and rural India today has more disposable income than urban India.

Rural marketing is getting new heights in addition to rural advertising. Rural marketing gives challenge to ensure availability of product or service in India's 6, 27000 villages spread over 3.2 million square kilometers. Marketers have to locate over 700 million rural Indian and finding them in not easy.

The size of the rural market is one that companies cannot afford to ignore, particular, as the number of simple lining in non-metro areas increased by 10 per cent over the past decade. Thus, looking at the challenges and opportunities, which rural markets offer to the marketers, it can be said that the future is very promising for these who can understand the dynamics of rural markets and export them to their best advantage.

MANAGE an extension management institution may provide extension services to rural public into information, price information, insurance, and credit information by using various media. "It is often said that markets are made not found. This is especially true for the rural market like India. Rural market is a market for a truly creative marketer. Civilization always begins with the development of villages, therefore, if needs high concentration - Mahatma Gandhi.

Rural marketing is currently growing at about 20 per cent every year and companies are spending amount ₹ 600 crore per years for promotional budget.

Nature and Importance

In the 21st century, the rural markets have acquired significance. The green revolution and the white revolution combined with the overall growth of Indian economy have resulted into substantial increase in the purchasing power of the rural communities. Rural marketing denotes blow of goods and services from rural producers to urban consumers at possible time with reasonable prices, and agricultural inputs and consumer goods from urban to rural. It is of paramount importance in the Indian marketing environment as rural and urban markets in India are so diverse in nature that urban marketing programmes just cannot be successfully extended to the rural market differs from that of the urban Indian. Further the values aspiration and needs of the rural people hasty differ from that of the urban population. Buying decisions are highly influenced by social customer's tradition and beliefs in the rural communities. As regards the purchasing power, the urban markets are segmented according to income levels, but in rural areas, the family incomes are grossly underestimated.

Farmers and rural artisans are paid in cash as well in kind, and their misrepresent their purchasing power. For their reason, a marketer must therefore, make an attempt to understand the rural consumer better before meaning any marketing plans.

Rural markets in India have untapped potential. There are several difficulties confronting the effort to fully explore the rural markets. The concept of rural markets in India is still in evolving shape, and the sector pages a variety of challenges. Distribution costs and non-availability of retail output are major problems faced by marketers.

Many successful brands have shown high note of failure in the rural markets because the marketers try to extend marketing plans that they use in urban areas. The unique consumption pattern, tastes, and need of the rural consumers should be analysed at the product planning stage so that they match the needs of the rural people.

Main Problems in Rural Marketing

Under Developed People and Underdeveloped Markets

The impact of agricultural technology is not felt uniformly throughout the country. Some districts in Punjab, Haryana and the Western U.P. where the rural consumers are somewhat comparable to their urban counter part; but there are large areas and grown of people who have repaired beyond the technological breakthrough. In addition, the farmers with small agricultural land holding are also unable to take advantage of the new technology.

Lack of Power Physical Communication Facilities

Nearly 50 per cent of the villages in India do not have all weather roads, physical communication to the villages is highly expensive. Especially during the monsoon 4 months these villages become complete inaccessible.

In Adequate Media Coverage for Rural-communication

A large number of rural families own radio and TV sets, there are also community radio and TV sets. These have been used to diffuse agricultural technology to rural areas. However, the coverage relating to marketing is inadequate.

Many Languages and Dialects

The number of languages and dialects vary from state to state and region to region. This type of distribution of population warrants appropriate strategies decide the extent of coverage of rural market.

Other Problems of Rural Marketing are Natural Calamites

Of draught or examine rain, epidemics, primitive methods of cultivation, lack of printer storage facilities, transportation problem and inadequate market intelligence, including long chain of intermediaries between cultivator and farmer and wholesaler and retailers.

There are also problems of extending marketing efforts to small villages with 200-500 population. Vast cultural diversity, vastly varying rural demographics, poor infrastructure, low income levels and low levels of literacy often tend to lower the presence of large companies in the rural markets.

Rural Marketing Strategy

Rural marketing strategy is based on their A's - Availability Affordability and Acceptability. The first 'A'-Availability emphasizes on the availability of the product for the customers, *i.e.,* this gives importance on effective distribution through efficient channels of distribution.

The second 'A'- Affordability which focuses on product pricing, i.e, this gives importance for smaller packages/pouches easily affordable by families in the rural areas, The third 'A' - Acceptability focuses on convincing the customers to buy the product, *i.e.,* extending suitable promotional efforts to influence the customers to buy the product. Marketers need to understand the psycho of the rural consumers and then act accordingly.

Rural marketing involves more intensive personal selling efforts compared to urban marketing. Firms should refrain from pushing goods designed for urban markets to the rural areas. To effectively tap the rural market a brand must associate it with the same things the rural consumers do.

This can be done by utilising the various rural folk media to reach them in their own language and in large number so that the brand can be associated with the myriad rituals, celebration, festivals, melas, fairs and weekly hats.

Rural Distribution Strategy

One of the ways would be using company delivery mass, which can serve two purposes - it can take the products to the customers in every hook and corner of the market and it also enables the firm to establish direct contact with them and thereby facilitate sales promotion. However, only the large manufactures can adopt this channel. The companies with relatively fewer resources can go in for the syndicated distribution where a tie-up between non-competitive marketers can be established to facilitate distribution.

Back-haul Method for the Distribution Vehicles

Organising a suitable back-hual method for distribution vehicles may prove to be an economic to transport the "urban goods" like soap, detergent, oil, cream, shampoo, tooth paste, and other daily necessary items for the rural consumers and in the return journey, the energy verticals will transport the fruit and vegetables etc. from rural areas to the nearest towns and cities for distribution among the urban consumers.

But this needs a well co-ordinated "VMS" distribution strategy in which the manufacturer, distributor/relation and the customers jointly make a strong distribution chain.

Annual "melas" and "fairs" organised are quite popular and provide a very good platform for distribution because profit visits them to make several purchases. According to the Indian Market Research (IMRB) Burean, around 8000 such nulas and fairs are held in the rural India every year.

Rural markets have the practice of faxing specific days in a week as weekly market days, *i.e.*, "Haats" when exchange of goods and services are carried out. This is another potential low cost distribution channel available for the marketers.

Also, every region consisting of several villages is generally served by one satellite town, formed as "Mandia" or Agri-markets where people prefer to go and buy from their commodities. The marketers using their feeder fown will be able to cover a large section of rural population.

The other distribution strategies for the rural population are as under:

- The general insurance companies may promote their policies of health insurance, crop insurance and vehicle insurance through the existing co-operatives.
- Marketers may arrange more number of wave-houses for storage and re-packaging into smaller pouches for which employing local villages will work profitable and popular.
- All communication in the rural areas must be in the regional language and dialects.
- Markets need to develop innovative packaging technology which would be economic, protective and improve shelf-life of goods.
- In addition to focusing on targeted promotions and advertising,

there is an urgent need to work on economical packaging, dual pricing and special size of PMCQ and household products.

- Marketers need to place emphasis on retailers directly rather than depending on the wholesalers for distribution in the rural market as this has not proved to be very effective marketing channel.
- Marketers targeting the rural market should be well aware about the seasonality of the business. Because the trade is seasonal, employment and disposable income can fluctuate arrange the villages during the year. This means that business should view market research data that relies on yearly aggregate statistics with caution.
- Marketers must trade off the distribution cost with incremental market penetration.

8

Promotional Management in Agriculture

Commercialisation of the agriculture has been the priority of various agricultural programmes at the recent times which has increased substantial marketable surplus of various agricultural commodities like vegetables, fruits, species, cash crops and other agricultural products within the country. With the increase in the volume of marketable surplus, the need for assured market outlets has become very necessary.

The nature of marketing communications

Not everyone believes that promotion is necessary. Both Marx and Lenin viewed advertising as a pernicious activity characteristic of bourgeois capitalism. Marx denounced advertising as "parasitic" whilst Lenin thought it irrelevant to socialism where centralised planning would ensure that exactly the right amount of product would be made available to meet consumer needs. It is hardly surprising, therefore, that advertising (excluding that taking the form of propaganda) has for a long time been restricted, controlled, and sometimes banned, in many of the nations which adopted communist and socialist political systems, including a good number of developing countries. Even after market liberalisation and political reform within these countries there often remains uncertainty over the need for advertising and other forms of promotion, and a suspicion that it adds nothing but costs to the marketing process.

In fact, without effective marketing communications the consumer remain unaware of products and services they need, who might supply them and the benefits which both product and suppliers can offer. Moreover, it is impossible to develop effective and efficient marketing systems without first establishing channels of communication. Even the best products do not sell themselves.

Marketing communications serve five key objectives:

- The provision of information
- The stimulation of demand
- Differentiating the product or service
- Underlining the product's value,
- Regulating sales.

Marketing communications takes four forms - advertising, sales promotion, personal selling and publicity. These must be formulated within a co-ordinated marketing communications plan. If there is more than one target market then there will need to be more than one communications programme. Like all other elements of the marketing mix, it must be tuned to the characteristics and needs of the target market.

Advertising: Advertising is the most visible element of the communications mix because it makes use of the mass media, *i.e.* newspapers, television, radio, magazines, bus hoardings and billboards. Mass consumption and geographically dispersed markets make advertising particularly appropriate for products that rely on sending the same promotional message to large audiences. Many of the objectives of advertising are only realised in the longer term and therefore it is largely a strategic marketing tool.

The objectives of advertising are broader than that of directly stimulating sales volumes. African Distillers, for example, contribute to a series of television advertisements, shown around the time of public holidays in Zimbabwe.

These warn people of the dangers and irresponsibility of driving when intoxicated. This involvement serves to enhance African Distillers' image as a socially responsible and caring organisation. The objective of this kind of advertising is to create a positive attitude towards the company on the part of its publics, *e.g.* government, pressure groups, shareholders, suppliers, agents and the general public. Some of these publics will never consume the company's products and this kind of advertising campaign is not intended to encourage them to do so.

Sales Promotion: Sales promotion employs short-term incentives, such as free gifts, money-off coupons, product samples etc., and its effects also tend to be short-term.

Therefore, sales promotion is a tactical marketing instrument. Sales promotions may be targetted either at consumers or members of the channel of distribution, or both.

Public relations: Public relations is an organisation's communications with its various publics. These publics include customers, suppliers, stockholders (shareholders, financial institutions and others with money invested in the business), employees, the government and the general public. In the past, organisations thought in terms of publicity rather than public relations. The distinction between advertising and publicity was based on whether or not payment was made to convey information via the mass media. Advertising requires payment by the sponsor of the message or information whilst publicity is information which the media decides to broadcast because it is considered newsworthy and therefore no payment is received by the media from a sponsor. It is more common these days to speak of public relations than of publicity. Public relations is much more focused in its purposes.

The objectives of public relations tend to be broader than those of other components of promotional strategy. It is concerned with the prestige and image of the organisation as a whole among groups whose attitudes and behaviour can impact upon the performance and aims of the organisation. To the extent that public relations is ever used in product promotion, it constitutes an indirect approach to promoting an organisations products and/or services.

Personal selling: This can be described as an interpersonal influence process involving an agribusiness' promotional presentation conducted on a person-to-person basis with the prospective buyer. It is used in both consumer and industrial marketing and is the dominant form of marketing communication in the case of the latter.

Developing an appropriate communications programme

Marketing strategy is derived from an organisation's corporate strategy. The marketing strategy then has to be translated into a strategic plan, or set of strategic plans if the organisation intends to exploit opportunities in more than one target market. Strategic plans are very broad statements of principles which the organisation believes will lead it to achieve its marketing objectives within a chosen target market.

These principles become operational when they are expressed in the form of a marketing plan consisting of a detailed blueprint for each element of the marketing mix product, distribution, pricing and marketing communications.

The connection between marketing communications and the marketing strategy. It also highlights the main stages involved in developing a marketing communications programme. The remainder of this chapter is devoted to explaining these stages.

Once the overall marketing strategy has been determined and the marketing plan has been outlined, it is necessary to develop a set of operational communication objectives. It is only when this is done that an appropriate marketing communications mix can be designed. There are, however, a number of intervening factors to be considered before the communications mix is finalised.

These include the nature of both the product and the market, the stage at which the product lies in its life cycle and the relative value of the product in terms of its price to potential purchasers. Having decided upon the communications mix, the promotional message can be determined and the medium or media best suited to delivering this message can be chosen.

At this point, the budgetary implications of the decisions made so far have to be considered.

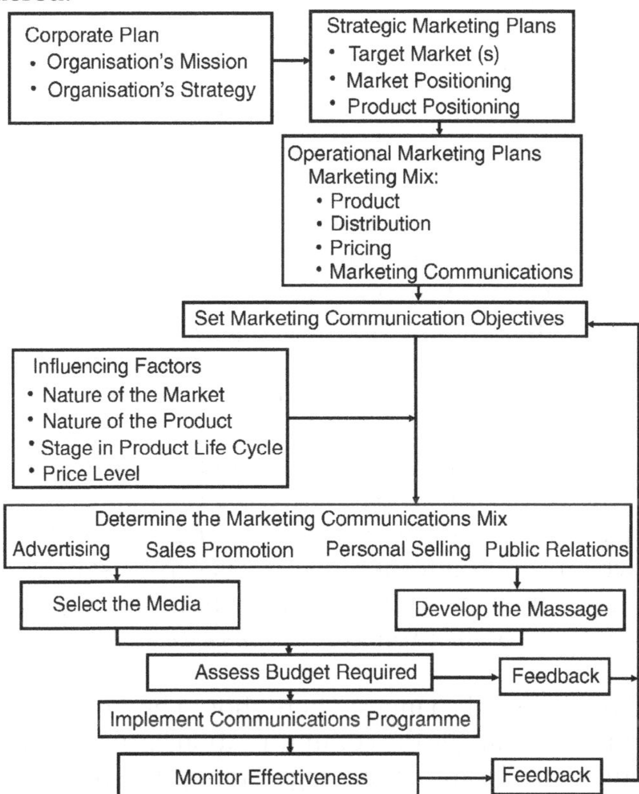

Fig. Developing a marketing communications programme.

If the cost of the communications programme exceeds the resources available to the organisation, then there may have to be an adjustment in the communications mix. In some instances, the organisation may conclude that it can adjust the communications mix to reduce the cost to an affordable level but that the revised communications package is unlikely to achieve the original objectives.

Faced with this situation, the organisation may resort to revising its marketing communications objectives. Once the budget has been set the programme can be implemented.

The effectiveness of the programme has to be measured against its objectives and, if necessary, adjustments or wholesale revisions of the programme will be made.

Setting marketing communication objectives

The question arises as to how operational communication objectives can be developed, given that these cannot be usefully defined in terms of sales volumes. A three step approach is proposed and this takes into account the longer term outcomes of marketing communications.

The three steps are:

- Identify the target segment
- Determine the behavioural change to be brought about
- Decide what needs to be done to bring about the change in behaviour.

Identifying the target segment: The identification of the target audience is obtained from the marketing strategy and marketing plan. There may, however, need to be a refinement of the target group for a particular promotional campaign. Returning to our earlier example of the problem of persuading farmers in the arid areas of Botswana to grow sorghum, it may be that the target group is defined as, "Those farmers operating small holdings of 5 hectares or less, in arid areas, who grew sorghum as a food crop in the past but stopped doing so completely to take up the growing of cash crops". There is a direct relationship between the degree of precision with which the target group is defined and the clarity with which communication objectives can be stated. Moreover, if the target group is defined with precision this greatly assists in deciding upon both the content of the promotional message and the medium chosen to carry it.

Intended behavioural changes: There should be a clear understanding of what behavioural changes the communications programme is intended to bring about. Is it: To increase usage among existing customers? To convert non-users to users? To establish new uses for an existing products? To reduce the amount of brand switching and encourage more users to be brand loyal? To enable customers to make better, more effective, more efficient or less wasteful use of the product and thereby increase its value to them? It is possible to measure the extent to which changes in behaviour have occurred, but marketing communication objectives can only become operational when the intended behavioural changes are stated with precision and without ambiguity.

Deciding what needs to be done: The third step in developing operation objectives for marketing communications, is to specify the required course of action. To increase the number of uses of a product might only require an awareness campaign, to improve the way in which a product is used (*e.g.* farmer's application of plant growth chemicals) would probably involve an educational campaign, to create a liking of the product a programme aimed at attitudinal change would be necessary, and the conversion of non-users of the product to users is likely to focus upon creating a conviction about its benefits and attributes.

Factors influencing the communications mix

There are at least 5 major influences on what makes a given mix of promotional techniques appropriate. These are: the nature of the market, the nature of the product, the stage in the product life cycle, price and the funds available for promotional activities.

Nature of the market: An organisation's target audience greatly influences the form of communication to be used. Where a market is comprised of relatively few buyers, in reasonable proximity to one another, then personal selling may prove efficient as well as effective. Conversely, large and dispersed markets are perhaps unsuitable for personal selling because the costs per contact will be high. The customer type also has an impact. A target market made up of industrial purchasers, wholesalers or retailers is more likely to be served by organisations which employ personal selling than is a market of consumers.

Another important consideration is the state of the prospective customer's knowledge and preferences with respect to the product or service. In some cases, the task will be to make potential customers aware of a product which is entirely new to them, whilst in others, the aim will be to attract them away from a competing product. The two tasks are quite different in nature and may require the use of differing forms of communication.

Nature of the product: Highly standardised products, with minimal servicing requirements, are less likely to depend upon personal selling than are custom designed products that are technically complex and/or require frequent servicing. Standardised, high sales volume products, especially consumer products, will probably rely more on advertising through the mass media.

Where the product is targetted at a narrow market segment or where those who can use the product effectively are few in number then personal selling will prove the more cost effective method of communications. For instance, in areas such as Pakistan and Sri Lanka, field sizes are too small for four wheeled tractors to work effectively. However, there may be a relatively small number of farmers who have larger fields and who can use such a tractor both effectively and efficiently. In these circumstances, a more direct approach to the target group of farmers would be advisable.

Stage in the product life cycle: The promotional mix must be matched to a product's stage in the product life cycle. During the introductory stage, heavy emphasis is placed upon personal selling to convey the attributes and benefits of the product.

Intermediaries are personally contacted to engender awareness, interest and, if possible, commitment to the product. Trade shows and demonstrations are also frequently used to inform and educate prospective dealers/retailers and, sometimes, consumers. Advertising at this stage is chiefly informative, and sales promotion techniques, such as product samples and money-off coupons, are designed to achieve the goals of getting potential customers to try the product.

As a product graduates into the growth and maturity stages, advertising places greater emphasis upon persuasion, with the ultimate objective of encouraging the target market to become purchasers of the product. Personal selling efforts continue to be directed at marketing intermediaries in an attempt to expand distribution. As more competitors enter the market, advertising begins to stress product differences to establish brand loyalty. Reminder advertisements begin to appear in the maturity and early decline stages. Thus, we see that as a product progresses through the product life cycle, both the marketing objectives, and the promotional mix used to achieve them, may well change.

Price: The fourth factor impinging upon the promotional mix is that of price. Advertising and/or sales promotion are the dominant promotional tools for low unit value products due to the high per contact costs in personal selling. Higher value products can justify, and usually require, personal selling.

Promotional budget: A real barrier to implementing any promotional strategy is the size of the promotional budget. Mass media advertising tends to be expensive although the message can reach large numbers of people and hence the cost per contact is relatively low. For many new or smaller firms the costs are prohibitive and they are forced to seek less efficient but cheaper methods. Ideally, a promotional strategy should first be developed and then costed rather than designing a promotional strategy around a preset budget.

Table. Choosing between personal selling and mass media.

	Personal Selling	Mass Media
Market		
Number of buyers	Few	Many
Geographic	Concentrated	Dispersed
Type of market	Industrial	Consumer
Product		
Product complexity	Custom	Standardised
Service level required	High	Low
Life cycle stage	Introductory to early growth	Maturity to early stage of decline
Pricing		
Budget	High unit value	Low value

The marketing communications mix

The next set of decisions is to determine the role of each element of the promotional mix.

Depending upon the situation, it is likely that more emphasis will be given to certain forms of promotion than to others. The main advantages and disadvantages of each element of the promotional mix.

Table. The main promotional methods.

Form of Promotion	Advantages	Disadvantages
Personal selling	Permits flexible presentation and gains immediate response.	Costs more than all other forms per contact. Difficult to attract good sales personnel.
Sales promotion	Gains attention and has instant effect.	Easy for others to imitate
Advertising	Appropriate for reaching mass audiences. Allows direct appeal and control over the message.	Considerable waste. Hard to demonstrate product. Hard to close sale. Difficult to measure results.
Public relations	Has a high degree of believability when done well.	Not as easily controlled as other forms. Difficult to demonstrate or measure results.

Advertising

Advertising is characterised as a form of communication which its sponsor pays to have transmitted via mass media such as television, radio, cinema screens, newspapers, magazines and direct mail. It is intended to both inform and persuade. Lancaster and Massingham describe advertising as being:

"...concerned with the identification and presentation of desirable and believable benefits to the target audience in the most cost effective way."

Table. Some of the aims of marketing communications.

Consumer Communications	Trade Communications	Corporate Communications[a]
Correct misconceptions about a product/service	Inform about promotional programmes	To establish, maintain or improve the corporate image
Increase frequency of use	Announce special trade offers	To inform its publics as to its values, policies and purposes
To remind of products or brands	To avoid stockpiling	To communicate its business performance
To present special offers	To build loyalty	To explain mergers and

		acquisitions to its publics
To educate on product usage	To educate on product usage	To explain fundamental changes in the organisation's mission.

Note:

[a]In some respects the aims of corporate communications would seem more a responsibility of public relations rather than advertising. However, reference here is made to the use of medium for which the organisation has paid. Public relations does not pay to make use of the mass media.

Harper believes that the same volume of advertising can have a greater effect in a developing country than it would have in a developed country because of the relatively low amount of advertising in LDCs and the low levels of competition between advertisers for the attention of audiences.

The aims of advertising are many. Some of the aims which advertising may be directed towards achieving.

Theory suggests that there is a lag between advertising activity and its effect on sales. Changes to consumer attitudes take time, as does creating customer awareness or creating an understanding of a product or product attributes. Thus, whilst advertising can undoubtedly have an immediate impact, the total effects of advertising are only realised in the longer term.

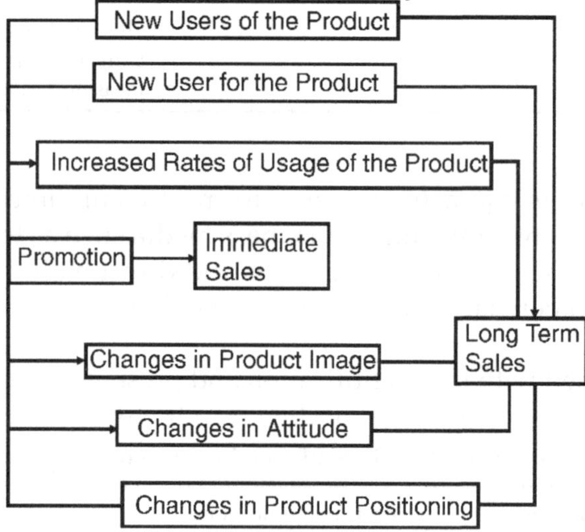

Fig. Longer term outcomes of promotion.

Since the effects of advertising are only evident in the longer term it should be treated as a strategic rather than tactical tool of the marketing communications mix. Advertising does not have the immediate impact of creating a customer. Instead, it has a hierarchy of effects. Lavidge and Steiner's hierarchy of effects model describes communication as a process rather than a simple outcome in the form of a sale.

Awareness: Consider the task facing a government which is attempting to persuade farmers in a frequently drought-stricken area to switch some of their production from maize to more drought-resistant sorghum. The initial step is for advertising to create an awareness of both the economic and technical benefits of sorghum which would accrue to farmers within drought-stricken areas. There may also be an awareness task to be accomplished with respect to new sorghum varieties whose higher yields help compensate for the superior economic rewards of growing maize in a good season. Levels of awareness can be measured and thereby used as a measure of the effectiveness of advertising. For example, prior to beginning a planned advertising campaign a target such as the following might be set:

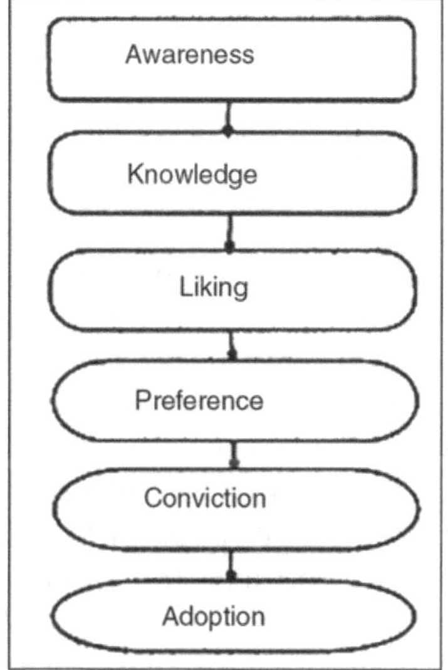

Fig. The hierarchy of effects model.

'Within 3 months of the campaign running, we expect at least 30 per cent of farmers in region X to be aware of the new sorghum variety and to be able to recall the 3 main technical benefits that are claimed for the variety in the campaign.'

Subsequent research among the region's farmers would permit management to determine whether the advertising had accomplished this target or not.

Knowledge: The next step is to instill, in farmers, a given level of knowledge about, for instance, how to choose economically viable sorghum varieties and the best husbandry practices to maximize yields; and economic results, the technical and commercial benefits of the new variety and how these are achieved. It is unlikely that advertising alone can communicate this type of information. The technical nature of the information would suggest that farmers would wish to put questions to sales personnel and/or extension agents in order to obtain further explanation.

Whatever combination of marketing communications is used, quantitative targets can again be set and the performance of the programme can be evaluated against them. It is particularly important that post-campaign research establishes the level of understanding among the target group. It should never be assumed that just because a message is received it is also understood.

Once an awareness and understanding has been built up among the target audience the marketer can then focus on establishing a liking or positive attitude towards the crop. This might be done, for instance, by promoting the virtues of the new variety, *e.g.* drought-resistant, high-yielding and palatable.

Preference: Even though the campaign may create a positive predisposition towards the product or service, the product may not be preferred to the alternatives. In the case of the hypothetical new sorghum variety, the target audience may like what it hears about the variety but this may not yet be preferred to existing varieties or to planting maize. Preference can be created by promoting the comparative advantages of the new product or service over its alternatives. In Botswana, the government was successful in promoting the production of sorghum by using the extension service to stress the versatility of the crop in use and, therefore, marketing opportunities. Sorghum can be made into soft porridge (motogo) or stiff porridge (papa or bogobe). It can also be used to produce three fermented products: traditional beer, "mageu", a non-alcoholic drink, and "ting", a fermented porridge. Whilst the versatility of the crop might create a preference for planting sorghum over maize in arid areas, additional benefits would have to exist in order to create a preference for this variety over other types of sorghum. Perhaps the new variety yields a particularly sour taste much favoured among drinkers of traditional beer in Botswana and/or is impregnated with queleatox to protect it from the main pest attacking sorghum, the quelea bird. To create preference the promotional message must convey benefits which alternatives do not possess.

Conviction: It is possible that whilst the target audience has developed a preference for a product or service their conviction about that product or service is not yet strong enough to actually cause them to adopt it. Here, the role of communication is to convince the target audience that the claimed benefits of the product or service are both real and sufficiently great to warrant a change in their behaviour. For example, prospective growers of the new sorghum variety will want to see the benefits for themselves through field trails and demonstration plots, and will perhaps want to converse with farmers who have already grown the new variety. This hypothetical example indicates that the medium of communication (*e.g.* printed media or demonstrations) and sources of information (*e.g.* extension personnel or other farmers) may change from stage to stage.

Adoption: The final step is for the target audience to adopt the crop, husbandry practice, product or service. The original hierarchy of effects model had purchase as its final step but here the term adoption is preferred because it emphasizes that the ultimate objective of promotion is to encourage a long term change in behaviour and not a one-off trial or purchase. To facilitate the initial purchase or trail of the product or service the promotional campaign might centre around a low introductory price or enable potential customers to try it on a limited basis. Prospective growers of the new sorghum variety could be offered seed at a discounted price or the seed might be specially packed in small sample sachets so that it could be sown on a trial plot of land. Target rates of adoption, over a specified time, should be set. However, as will be explained a little later in this chapter, direct comparisons between the number of adopters and promotional activity would not be meaningful since there are many other intervening variable. If targets are not being met, then what can be done is to reassess promotional efforts.

In particular, research needs to be carried out to determine answers to the following questions:

- Is the unique selling proposition (USP) understood and valued? (*e.g.* growers may not understand how queleatox works and will therefore have difficulty accepting its benefits or may believe that other pests such as grasshoppers and locusts are more of a threat to the crop).
- Was the right communication medium used? (*e.g.* newspapers, magazines and leaflets may have been used where word-of-mouth communication through sales personnel or extension agents might be more effective in communicating the complexities of sorghum growing and/or marketing. In many instances the mass media is effective in creating awareness, interest and communicating information but personal communication is required to effect trial and adoption).
- Did the message reach the intended audience? (*e.g.* it might be established that the majority of prospective sorghum growers listen to a given radio station but the message is not transmitted at peak listening times for this group).
- Was the source of information acceptable? (*e.g.* farmers may suspect

that the government is motivated by a desire to curtail over-production of maize rather than to reduce the risks of small-holders farming in arid areas whereas the same message might be more readily accepted if the source were an independent research station.

Put another way, the key questions are:

Message: Is the right message being communicated?

Media: Is the right medium or media being employed?

Target: Is the target being reached by the communication?

Source: Is the source of the information credible with the target audience?

There is a great deal of empirical evidence to support the notion that there are distinct stages in the communication process and also that the effects of this process occur over time. Studies by Beal and Rogers into the adoption of herbicides and new livestock feed formulations by farmers showed not only the distinct steps in the communication process described here as the hierarchy-of-effects, but also that lapses of several years between awareness and adoption can occur. Similar evidence has been provided by Singh and Pareek from their studies of farmers in India.

In summary, what needs to be recognised is that it is unlikely that all of the steps in the communication process can be accomplished by a single advertisement or advertising campaign. The first two steps, for instance, 'creating awareness' and 'developing knowledge', differ in that the first merely requires that the audience be informed by reaching them whilst the second demands that they be educated. The two tasks are quite different and are, invariably, achieved in different ways. Similarly, the subsequent tasks of creating, a positive predisposition, preference (or loyalty) and adoption of the idea, product or service are different in nature and are most likely to differ in method.

Rogers suggests that:

- "Mass media channels are relatively more important at the knowledge stage and interpersonal channels are relatively more important at the persuasion stage...".

It should also be recognised that since promotion has a number of intermediary goals its performance cannot be measured simply in terms of sales volumes.

Given that many of the outcomes of advertising are realised only in the longer term, there is little value in attempting to set advertising objectives in terms of immediate sales because there is little prospect of being able to directly associate a given rise (or fall) in sales with a particular promotional campaign or component of a particular campaign. This being the case, promotional objectives based on sales effects do not meet Aaker and Myers' requirement that advertising objectives be operational because they do not provide decision makers with useful criteria on which to base future decisions.

Transforming non-buyers into buyers is seldom achieved in a single direct step. Instead, advertising seeks to take prospective customer through a series of distinct steps. Whereas it is not possible to measure the impact of promotional activities in terms of sales effects, targets can be set for each of the intermediate goals which comprise the hierarchy of effects model depicted below. Research can be conducted subsequently to determine the extent to which these goals have been achieved.

Aaker and Myers say that:

"Advertising objectives, like organisational objectives, should be operational. They should be effective criteria for decision making and should provide standards with which results can be compared. Furthermore, they should be effective communication tools, providing a line between strategic and tactical decisions."

It might be thought that the primary objective set for advertising would relate either to immediate sales or to market share targets, but in practice such objectives are seldom, if ever, operational.

The reasons are threefold:

- Advertising is only one of many factors influencing sales
- The effects of advertising are often long term and
- The effects of advertising are usually indirect.

Some of the many factors which affect sales levels and make it difficult to isolate the effects of advertising. There is a dichotomy of factors, these being the endogenous factors operating from within the individual, such as his/her attitudes, opinions etc., and those external or exogenous to the individual, including the elements of the marketing mix, the general economic climate and cultural influences.

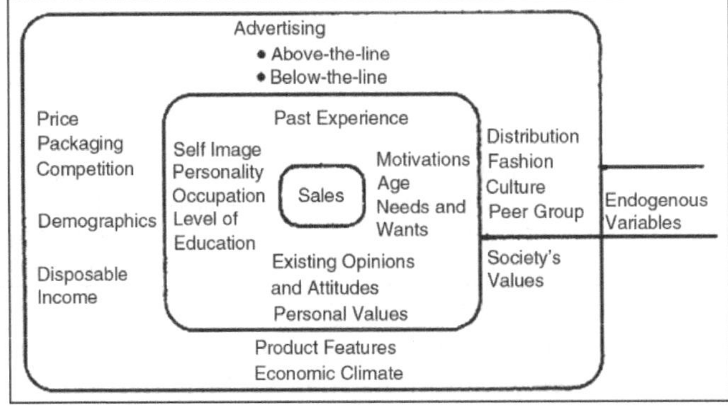

Fig. Some of the factors influencing sales.

Sales promotion

In contrast to advertising, sales promotion is more tactical than strategic. It is usually applied to create an immediate impact, but one which is unlikely to be sustained in the longer term. Thus, marketers tend to use promotion to address short term problems such as reducing the cash burden of overstocked products, stimulating demand during what is traditionally the low season, selling off stocks which are becoming obsolete or are likely to spoil if they remain in storage.

Sales promotions may be targetted at consumers, industrial buyers (*e.g.* crop processors or food manufacturers), channel intermediaries (*e.g.* traders, wholesalers or retailers) or the organisation's own sales force.

Table. Types of consumer sales promotion.

	Sales Promotions Targetted On Customers
Type of promotion	**Examples**
Discount coupons or money-off packs	Discounts on the full price encourage product trial, *e.g.* $5 off the regular price that will apply to a new pesticide.
Premiums	Products offered free or at a discount act as an incentive to buy a related product, *e.g.* farmers buying 25 litres or more of a new pesticide get a 5 litre pack of herbicide free.
Lotteries, games or competitions	Intended to create interest and excitement among customers, *e.g.* farmers may be offered the opportunity to win a knapsack or tractor mounted agrochemical sprayer.
Samples	Free samples encourage product trial, *e.g.* farmers could be given a small pack of pesticide and invited to apply it to a small plot and compare results either with a plot to which no pesticide has been applied or against a competing brand.
Point-of-sale merchandising	These specially designed display units and literature are intended to create impulse (*i.e.* unplanned) purchases. They are located close to the place where the customer pays for the goods or service, *e.g.* the packs of pesticide could be arranged on an attractive rack displaying the manufacturer's name and situated, close to the checkout in a farmers service centre.
Trading stamps	Customers are given stamps in ratio to the value of their purchases. Each stamp has a value attached to it although it cannot be redeemed for cash. These stamps can be accumulated and then traded in as whole or partial payment for goods and services. Trading stamps are mainly used by distributive outlets to encourage customer loyalty.

Table above gives examples of typical forms which sales promotion takes. Many of these forms are equally applicable in consumer and industrial markets. Sales promotions may be targetted on intermediaries as well, or instead of, consumers. Many types of promotion are used in both sectors. Sometimes, however, their objectives are slightly different.

Table below describes the main forms of trade promotions and their various purposes.

Table. Types of trade sales promotions.

	Sales Promotions Targetted On Trade Channels
Type of promotion	**Examples**
Trade allowances	These temporary price reductions are intended to be passed on, in whole or in part, to the end customer. Thus, intermediaries can elect to have a higher margin per unit or higher volume sales.
Bonus purchases	An agricultural merchant may be offered 24 packs of pesticide for the price of 20. Such bonuses are not intended to be passed on to customers but are an incentive for the merchant to increase the order size.
Competitions	These are directed at the sales and/or service personnel of intermediaries and if sponsored by a manufacturer/grower are intended as an incentive to place particular emphasis on selling that supplier's products or services., *e.g.* a salesman achieving total orders in excess of 1,000 litres of pesticide might win a cash prize.
Cash incentives	Cash bonuses paid to a middleman's sales personnel can help push the product through the channel of distribution.
Cooperative advertising/promotions	Suppliers and middlemen sometimes share the cost of an advertising campaign or promotion, *e.g.* An agricultural merchant wishing to run a local campaign may obtain assistance from one or other of his/her main suppliers.
Trade shows and exhibitions	An industry's trade association may organise fairs and exhibitions which offer its members the opportunity to communicate with a well defined target audience. Both manufacturers and intermediaries may participate in these events.

Public relations

Publicity and public relations are not one and the same thing. Organisations often seek publicity, *i.e.* to disseminate newsworthy items of information about itself, its products/services or about its personnel through the media but does not pay to do so as in the case of advertising. Instead, the organisation hopes that the item is sufficiently newsworthy to appear in an editorial feature, in a newspaper or magazine, or that a radio and/or television station will want to interview an official of the organisation about the item.

Publicity can be a highly effective communication tool, since 'news' is often perceived by the target group to have greater authenticity and credibility than 'advertising'. Moreover, it can penetrate the defences of individuals who intentionally ignore advertising and the overtures of sales personnel. The main disadvantage of publicity is that the organisation has relatively little control over it.

By contrast, the organisation can exert a large degree of control over the results of public relations so long as there are specific objectives in regard to each of the publics at which the communications are to be directed.

Public relations may be defined as:

"...the deliberate, planned and sustained effort to establish and maintain mutual understanding between an organisation and its public."

The 'public' referred to in this definition is any group having an actual or potential interest in, or impact upon, an organisation's prospects of achieving its goals. Such publics would be:

The community: An organisation needs to be accepted by the local community. To this end, a community relations programme should be established. Such a programme should devise ways for the organisation to become involved in community activities. A public relations programme can give an organisation a 'personality' and, hopefully, one which the local community likes.

Consumers: Public relations should be used to nurture a positive image of the organisation and its products and services, a belief in its intrinsic fairness in dealings with customers and the perception that the organisation values loyal customers.

Other channel members: Wherever the organisation is placed within the marketing channel (as a grower, processor, wholesaler, retailer etc.) it should take cognisance of the need to develop and maintain positive relations with its partners within the marketing system. The public relations programme should make them feel like partners, *e.g.* by making them privy to privileged information about the organisation's products, promotional programmes, marketing plans, future developments and/or policies.

Opinion leaders: Pressure groups and trade associations are examples of groups which can influence both public and government opinion and therefore should be a target for the organisation's public relations activities Where there is potential conflict between the interests of these groups and those of the organisation it is vital that there remains a dialogue between them so that factual information, rather than rumours, is communicated. In many cases, an effective public relations programme can help avoid conflicts from arising.

It can do so by projecting a corporate image of a caring, responsible and responsive organisation. For its part, the organisation must seek to understand the position taken by pressure groups on particular issues.

Government: The lobbying of politicians is a sensitive issue but in most countries around the world it is accepted as a reality. Public relations programmes should be designed to create a two-way flow of communications between industry and government (or between a trade association, such as a farmers' union, and government). That is, the organisation should be creating a positive predisposition towards it whilst it should be receiving advance information, from government, on matters such as proposed legislative changes that could impact upon its activities.

Financial institutions: Bankers, finance brokers, investment analysts and other lending institutions are an important public for all commercial organisations. They need to have confidence in the financial stability and prospects for growth since directly or indirectly they will affect the organisation's access to debt capital. Public relations programmes targetted at this group are therefore very important.

Media: Sound press relations can give an organisation access to the 'news' channel of communication through which it can disseminate positive information to all of its publics. Through its public relations programme, the press should be given a ready response to all reasonable requests for information within the limits of commercial confidentiality, that the organisation is candid about its intentions and actions.

Employees: Organisations must recognise the need to 'market' themselves to their own employees as much as to other publics. Internal public relations often suffers from neglect. The loyalty and commitment of employees to the organisation and its goals cannot be taken for granted. An internal public relations programme can also help build an understanding between the organisation and its personnel as well as helping develop an enduring trust between them.

The methods employed by public relations professionals include:

- Open days
- Sponsorship
- In-house publications
- Community projects
- Press releases
- Video films
- Training courses,
- Annual reports.

Public relations has perhaps a different but complementary role to that of other forms of communication. It will be most effective, and controllable, when it has specific objectives, with respect to specific publics, and when it is coordinated with the forms of marketing communication.

Personal selling

Personal selling complements both advertising and sales promotion. Many organisations have a sales force comprised of a number of representatives who have face-to-face contact with the customer. The division of responsibilities between sales representatives may be based on geographical areas or on product groups. For instance, an agrochemical company could divide the market into geographic regions and assign a representative to each district.

He or she would have responsibility for selling all of the company's products to the assigned area. Alternatively, the same agrochemical company could organise its sales force so that representatives handle either animal health products or crop protection products. This would make sense if, within a country, farming tended to be specialised into arable and livestock, whereas it would perhaps be less appropriate if mixed farming were the norm and two representatives, from the same firm, were calling on the same farmer.

Reid defines personal selling as:

"...the process of analysing potential customers' needs and wants and assisting them in discovering how such needs and wants can best be satisfied by the purchase of a specific product, service or idea."

Given the importance of personal selling within the marketing mix and the fact that the sales force is likely to be the most expensive element of the company's promotional mix, the organisation must be clear on the objectives of its sales representatives.

Sales representatives have at least 7 key tasks:

Prospecting	Sales representatives find and develop business with new customers.
Communicating	Sales representatives communicate information about the organisation's philosophy, produce/products and/or services and communicate needs. preferences and problems which customers have and the organisation can meet or resolve.
Selling	Sales representatives should be trained in the art of selling approaching, presenting, countering objections, closing sales and nurturing a long-term customer relationship.
Servicing	Sales representatives provide various services to customers, such as helping resolve their problems with his/her own organisation, rendering technical assistance, arranging financing and expediting delivery.
Information gathering	Sales representatives carry out market research and intelligence work and complete visit reports. Representatives are able to collect information on competitor activity as well as the future needs of customers.
Complementing advertising	The activities of sales representatives should complement other elements of the promotional mix. The sales approach has to be consistent with the selling propositions conveyed through advertising and sales promotion.

	Where possible, customer visits should be timed to coordinate with the other promotional mix components.
Allocating	Sales representatives are able to evaluate the value of various customers to the organisation and advise on the allocation of scarce produce/products at times of shortage.

Thus, we observe that whilst selling is of fundamental importance, the sales representative has a number of other vital objectives, but at core he/she is part the organisation's promotional effort and is an important contributor to marketing communications.

In practice, companies will be more specific about how they expect their sales representatives to spend their time. For example, sales personnel may be told what proportion of their time to devote to existing products and customers and how much to spend on prospecting for new business or developing markets for new products.

Left to decide for themselves, sales people are likely to devote much of their time to exisitng customers where they know what kind of reception awaits them and to products they are familiar with and which have an established market (especially where sales commission is paid). The sales manager who permits this pattern to emerge is clearly unaware of the concept of product life cycles and the dangers of relying entirely upon today's customers and today's products.

One would expect to observe a difference in the objectives set for the sales force in a market oriented versus a selling/production oriented organisation. In the case of the latter, the accent is wholly upon sales volume and the sales force has no role to play in marketing strategy or issues relating to profitability.

A contrasting view should be in evidence where selling is perceived by management to be the central activity within the promotional element of the marketing mix.

The market oriented company trains its sales force to produce customer satisfaction and company profit. This involves developing the analytical marketing skills of sales personnel.

Training the sales force

Organisations which send their newly hired sales representatives immediately into the field are almost invariably disappointed by the results. It is true that training programmes can be expensive. Trainers have to be hired, materials purchased and, perhaps, facilities have to be rented. Moreover, the organisations are paying people who are not yet selling. There are also opportunity costs. Experienced sales representatives have to be withdrawn from the field for on-going training, and sales opportunities may be foregone. However, training, and re-training, is necessary if the sales force is to be effective. The sales manager's task is to ensure that training costs are outweighed by the value added to the company's business by having a better equipped sales force.

Not all sales personnel sell: many could better be described as order takers. This is not necessarily a problem as long as this is the purpose for which the "sales force" was intended. Many consumer food products are presold through extensive advertising in the mass media. In these circumstances, the sales person's role is largely confined to taking orders from middlemen who are already motivated to stock the product because of the demand created (or maintained) by the mass media. However, it becomes a problem when an organisation requires its sales force to take a more proactive role and aggressively sell its products and services. Despite stories to the contrary, there are very few 'born salesmen'. The great majority of sales personnel need to be trained to become active sellers as opposed to being passive order takers.

Sales force training programmes have several goals, including:

Sales representatives need to understand and identify with the company.	The first part of most sales training programmes is devoted to describing the company's history and objectives, the organisation and lines of authority, the chief officers, the company's financial structure and facilities, and the chief products and sales volume, as well as the current and prospective customer base.
Sales representatives need product knowledge.	Sales trainees are shown how the products are produced and how they function in various uses and in different environments.
Sales representatives need to understand customers' needs, buying motives, and buying habits.	They need to learn about the organisation's and competitors' strategies and policies.
Sales representatives need to know how to make effective sales presentations.	Sales representatives require training in the principles of salesmanship. In addition, the company outlines the major sales arguments for each product, and some provide a sales script.
Sales representatives need to understand field procedures and responsibilities.	Sales representatives learn how to divide their time between existing and prospective customers; how to prepare reports, organise their schedules and select efficient routes.
Sales representatives need to understand their role in marketing intelligence gathering.	Some individuals are quicker than others to realise that they have a role in market intelligence gathering. Indeed, representatives repeatedly fail to report information collected in the course of their work because they do not appreciate that it constitutes marketing intelligence. Hence, the need for training in this area.

Training programmes have to be evaluated against the performance (and/or improved performance) of the sales force. Quantitative evidence might include increased sales turnover and sales volumes, larger average order sizes, increases in new accounts, a decline in customer complaints and returns, lower levels of absenteeism, etc.

Change agents

Sales representatives are agents of change. They seek to change customer behaviour in many different ways: from mechanical weeding to herbicides, from bulk transportation of vegetables to prepacking, from broadcasting of seed to the use of precision drills, from employing manual labour to electro-mechanical grain elevators, from open to refrigerated lorries, from purchasing 10 kg bags of maize meal to buying 25 kg sacks, and so on. However, sales representatives are not the only agents of change seeking to communicate with producers, processors, traders, retailers, consumers, and other participants in agricultural and food marketing systems, in an attempt to alter their beliefs, opinions and behaviour in some way.

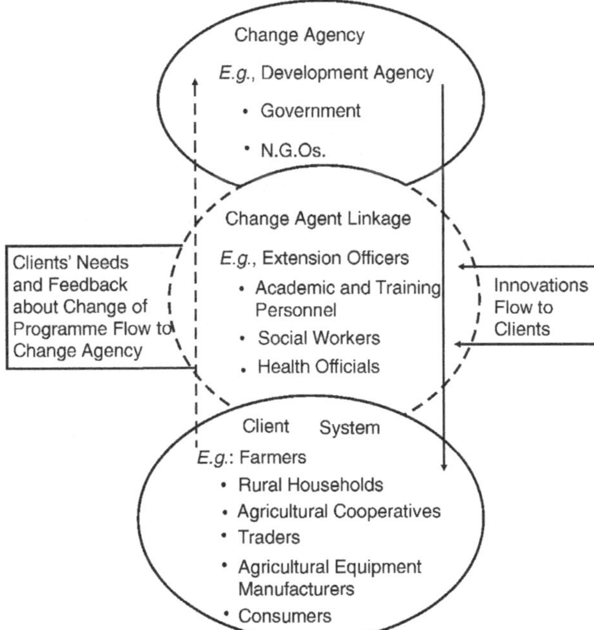

Fig. Change agents and their role in the context of agricultural and food marketing systems.

Agricultural extension agents, agricultural marketing extension agents, health workers, farmers' unions, cooperatives, trade associations, professional bodies, consumer associations and NGOs of various types are some of the other change agents whose activities are often found to impinge upon the agricultural and food marketing systems. For example, where change agents are successful in persuading farmers to adopt improved husbandry practices or technological innovations which lead to higher levels of production and/or crop quality, this is likely to have an impact in the marketplace.

Similarly, if change agents persuaded Southern African consumers to eat yellow maize (often considered food for livestock) in the place of the traditionally preferred white maize, then this would have an impact on the relative prices of the two grains and upon future levels of plantings and supply of white and yellow maize.

The figure overleaf depicts the role of change agents in the context of agricultural and food marketing systems. It shows the role of change agents, whether they be commercial sales representatives or other types of change agent, is that of linkage between the promoters of change and those who are expected to adopt and implement those changes. The types of clients whose beliefs, attitudes, opinions and behaviour change agents might seek to alter.

The change agent forms the link between the change agency and the target client group(s). He/she is a conduit through which information flows, in both directions, between the change agency and the client group. The change agent interprets and transmits the message of change from the change agency to the clients. The change agent is equally responsible for interpreting and transmitting information on the needs and problems of the clients to the change agency, also their reactions to proposed changes, obstacles, difficulties, and the identity of important opinion leaders within the client system.

The objectives of change agents have a particular sequence, and may be explained as follows:

1. Develops the need for change.	A change agent's initial objective is often to help clients see the need for change. For example, an extension officer might have to demonstrate how the lack of coordination between farmers in their production and lack of cooperation in the marketing of their produce prevents them from gaining access to distant markets and obtaining reasonable prices.
2. Establish a rapport with the client group.	Before they will accept his/her advice, the client group needs to feel that the change agent has empathy with their situation. A frequent obstacle to rapport is where the change agent comes from a different culture, does not speak the language (or dialect), comes from a different socio-economic class or intimidates the client group through ostentatious show of a superior education.

3. Diagnoses of the client groups' problems.	Returning to the earlier example, the change agent will have determined, through diagnoses, why the existing marketing system is not working and what the alternative solutions might be.
4. Encourages an intention to change among the client group	Having explored various alternative course of action, the change agent must motivate the clients to adopt one or other solution. This might involve sending trial shipments of produce to market, for example, to demonstrate that graded produce, carefully packed, earns higher returns for farmers.
5. Translates intent into action.	The change agent induces client-centred change. He/she will work through local leaders and opinion leaders so that the proposed change is adopted by the clients rather than always being seen as a solution imposed by the change agency. He/she may, for instance, encourage opinion leaders to call for farmers meetings to discuss how local production might be coordinated and which marketing functions should be based on cooperative activity.
6. Firmly roots adoption and minimizes the likelihood of its abandonment.	Change agents will seek to reinforce the new behaviour by continual feedback on improvements and benefits.This might take the form of reports on reductions in produce losses due to improved handling and packaging, higher prices than pre-change prices, increases in sales volumes, etc.
7. Makes the change self-perpetuating.	Eventually, the change agent should make him/herself redundant. The client group should become self-reliant rather than continuing to rely upon the change agent. This might be evidenced, for example, if farmers not only continued to coordinate their existing production and cooperate in the marketing of their produce but also began to look for better methods and new areas of cooperation in production and marketing.

Whilst sales representatives and other types of change agent are similar in many ways, they are not the same. Sales representatives are essentially profit oriented, as they must be, and this governs their selection of priorities and their behaviour. Other types of change agent are usually, if not always, motivated by social goals. Even when their efforts are directed at improving the economic performance of a client group, this is normally a means to an end and not the end itself. That 'end' is usually the economic and social development of the client group.

Developing the message

In most instances, the attention of the target audience can only be held for a relatively short time. That is, the potential customer will spend only a matter of seconds, or at most minutes, reading an advertisement in the printed media, will spare a limited time conversing with sales personnel or extension agents and will quickly lose interest in broadcast messages when these are perceived to be too long. Thus marketers must be selective in the points of information they seek to communicate. Whilst a product or service could have a large set of selling points, these will have to be narrowed down to a select few. Moreover, the single most important selling point will be the one to be included in the principal slogan or headline. This is sometimes termed the unique selling proposition (USP). A USP should only be decided upon after customer research has determined meaningful and important messages (*e.g.* there is little to be gained from promoting the nutritional superiority of hammer milled whole grain over roller milled refined grain when consumers believe the latter is superior in taste, colour and texture).

Selecting the media

The media plan has to be developed in concert with the overall marketing communications strategy. The hierarchy of effects model, described earlier in this chapter, stressed the multiple stages through which the target customer must be brought and that different media might be more successful at some stages than others. Therefore, it is likely that a mix of media will have to be used within a single marketing communications programme.

Criteria for selecting communications media include:

- Level of exposure
- Level of impact
- Nature of the target audience
- Cost and cost effectiveness.

Message exposure: Marketers are interested in the potential number of message exposures that a given medium offers. The total level of exposure is a function of reach and frequency. Reach is the number of people exposed to the message. For example, to the extent that a higher percentage of rural populations, in developing countries, have access to radio as opposed to television, radio will have the greater reach for this target audience. Frequency is the average number of times an individual is exposed to the message. If the target audience were say farmers, who tend to read a newspaper 2 - 3 times per week but listen to the farming news, on radio, 7 days per week, then radio is likely to achieve the higher frequency rating.

Invariably, there is a trade-off between reach and frequency. Communications budgets will stretch only so far and so more spent on one will reduce the amount that can be spent on the other.

Impact of the promotional message: It can be argued that the impact of a promotion has more to do with the message than the medium. Nonetheless, the medium is an influencing factor on the levels of awareness, comprehension, believability and retention. Radio, being a purely audio medium, will be limited in its impact on farmers' levels of understanding of the operation of a piece of agricultural equipment that is new to them. Visual communication would be important in this case. In the same way, the retention of information is generally higher for audio-visual communications than it is when the information is presented only in audio form.

The target audience: Media have to be selected according to their ability to reach the target audience. This involves analysing the demographic structure of the market socio-economic groups, age groups, language, ethnic groups, etc. Thereafter, marketers can seek to identify media that reach the target group(s).

Cost and cost effectiveness: Some forms of media may prove too expensive for a particular communications budget and although these may have great potential in reaching the target audience, they will be unavailable. Even when this is not the case, it is incumbent upon the marketer to identify the most cost effective media.

The cost-per-thousand method (CPM) is one of the most commonly used in measuring the cost effectiveness of promotional media. For example, if it costs \$100,000 to send a mobile cinema around the rural areas for 6 months, to demonstrate the advantages of applying herbicide, and if it is estimated that some 50,000 farmers will see the cine/video film, then the cost per thousand is:

$$\frac{\text{Cost} \times 1{,}000}{\text{Exposure to target Group}} = \frac{\$100{,}000 \times 1{,}000}{50{,}000} = \$200$$

The same calculation can be undertaken for alternative media which are also under consideration.

However, when making comparisons of this kind, care has to be taken to allow for the precision of a medium in hitting the target, something which the CPM approach does not do.

For instance, in some African countries fish trading has traditionally been carried out by people of Asian origin. If these were the target group for a given promotion then an Asian language newspaper might give pinpoint accuracy in reaching them but would score badly on a CPM rating since they are a minority of the population.

Establishing the promotional budget

Deciding upon the amount to be spent on promotion is one of the most challenging tasks marketing managers have to face. There are simply no scientific solutions to the problem. Since no one has ever established a mathematical relationship between promotional expenditures and their effects, either in terms of sales volumes or revenues, there is no universally accepted formula for setting the promotional budget. Instead, a number of pragmatic approaches have been established over the years. The main budget setting methods are percentage-of-sales, fixed-sum-per-unit, competitive parity, residual-sum and objective-and-task.

Percentage-of-sales: The method involves setting the budget as a percentage of either last year's sales or forecasted sales for next year. Thus, brands or products which are performing well get additional promotional support. The popularity of this approach is probably due to its simplicity. However, it suffers from several weaknesses, for example high sales volumes do not necessarily reflect high profitability, there is little support for marketing managers wanting to turn 'problem' products into 'rising stars' and when budgets are set according to forecasted sales there is motivation to inflate sales estimates. Another problem with this approach is that using percentages of sales leads to sales determining the promotional mix. It was suggested that the reverse was the correct relationship, *i.e.* the promotional mix should determine sales.

Fixed-sum-per-unit: Some organisations elect to set a specified amount for each unit sold or produced. Thus, for example, a poultry producers might determine the promotional budget by using a figure of $1.50 per gross of eggs sold and 25 cents per broiler sold (or produced).

Competitive parity: This approach is simply one of keeping pace with immediate competitors. The organisation will try to work out approximate expenditure levels by two or three close competitors and will then seek to match those expenditures. It represents a reactive or defensive approach to promotional budget setting. Apart from the difficulties of arriving at reasonably accurate estimates of expenditure by the competition, the method suffers from incorporating the mistakes of competitors who may be spending too much or too little. Alternatively, the amount spent by the competition might be right for them but not for others who have different resource levels and marketing opportunities or problems. The method also discourages organisations from taking a more aggressive marketing stance by seeking to gain a competitive advantage.

Residual-sum: This is a euphemistic term for allotting what the organisation perceives itself to be able to afford after all other budgets have been set. The danger is that in years of good business there may be over-budgeting whilst in times of low sales, when demand most needs to be stimulated, the amount available for promotion falls.

Objective-and-task: An organisation employing the objective-task approach will first specify its communication objectives and will then estimate how much it will cost to achieve those objectives. This is the approach to promotional budget setting recommended in this text. It has the benefit of encouraging marketing managers to set specific communication goals. When these are not attained the communications mix can be reevaluated and modified.

It may be that whilst the communication objectives are valid, the particular organisation cannot supply the resources to meet them. In these circumstances some sort of compromise between expenditure and goals will be necessary.

Perhaps the fundamental weakness of this budgeting method is the implied assumption of cause-and-effect. That is, there is an assumption of a direct relationship between promotion and marketing performance but as has already been said, other elements of the marketing mix will impact upon sales, as will many uncontrollable exogenous factors.

Whichever approach to setting the promotional budget is chosen it should be recognised that it has been established on a less than optimal basis.

Monitoring the effectiveness of marketing communications

The last step in the development of a marketing communications programme is to evaluate the effectiveness of the programme. The evaluation has two components: communications effects and market performance.

Communications effects: Research into communications effects involves the evaluation of a single advertisement. Research in this area focuses upon measuring variables such as attention levels, message comprehension, message retention and intention to purchase. This type of research is often termed copy research. Both broadcast and printed promotional material can be evaluated. Whilst the techniques employed differ in their detail they essential involve exposing a sample of people drawn from the intended target group and exposing them to the proposed advertisements or promotions. For example, a printed advertisement can be inserted in a dummy magazine and given to the sample. After a suitable period of time these people are asked to recall the advertisements seen and to report as much of the detail of the content of the ads selling propositions, images, applications, etc. In the same way, an audience can be recruited to watch television programmes with trial advertisements inserted at the beginning of the programme(s) and/or in the commercial intervals and/or at the end of the programmes. They too can be questioned about the content of the advertisements and the impressions that they made upon the audience.

Market Performance: To a limited extent and in certain situations, the effects of promotion on sales can be measured. The effects of special offers and coupons can be measured by redemption rates.

Two approaches which are widely pursued in industrialised countries are as follows:

Field experiments.	The organisation selects two geographical areas which are similar in terms of socio-economic groups, levels of disposable income etc. and launches a promotional campaign in one area but not the other. After a period of time, sales in the two areas are compared. The assumption is that the only difference between the two areas is the absence or presence of the promotional campaign and so differences in sales are explained by the promotional campaign.
Analysis of historical market data.	Promotional expenditures and sales data can be compared using mathematical or econometric models to first describe the relationship between sales and promotion. Where these can be successfully described there is the prospect of developing other models capable of predicting sales, given a certain level of promotional effort.

The first of these approaches requires the application of very strict controls and careful matching of the areas or markets to be compared. The second approach requires good quality data. That is, the data must be detailed, precise, free from error and must extend over a considerable period of time.

Bibliography

A. Poshadri and Aparna Kuna : *A Handbook of Food Techno's*, New India Publishing Agency, Delhi, 2013.

A.D. Dholakia: *Delicious Seafood Recipes*, Daya Publications, Delhi, 2010.

Andrew L. Winton and Kate Barber Winton : *A Handbook of Structure and Composition of Foods*, Agrobios Publication, Delhi, 2013.

Ashok K Chauhan and Ajit Varma:*Microbes : Health and Environment*, I K International, Delhi, 2007.

Aysha Aziz Faridi:*Dairy Microbiology*, Random Publications, Delhi, 2012.

Cristobal Noe Aguilar; Juliana Morales Castro; Efren Delgado; Diana Jasso Cantu and Ashok Pandey: *Food Science and Food*

Dalip Kumar and Asmi Raza: *Agriculture and Food Security: Contemporary Issues*, Deep and Deep Publications, Delhi, 2011.

Doreen Virtue : *Constant Craving : What Your Food Cravings Mean And How To Overcome Them* , Hay House India, Delhi, 2011.

Harmeet Singh: *Dairy Farming*, APH Publication, Delhi, 2011.

Hema Thapar: *Food Science and Health*, Pacific Publications, Delhi, 2011.

Hema Thapar: *Nutrition and Food Science*, Pacific Books International, Delhi, 2011.

M Lakshmi Narasaiah: *Economic Growth and Food Security*, Discovery Publishing House, Delhi, 2008.

Madhulika Bhatnagar.: *Encyclopaedia of Catering Technology, Food Service and Hospitality Management, Vol. I to III*, Delhi, Anmol Publication, 2008.

Manju Malhi: *Easy Indian Cookbook: The Step-By-Step Guide to Deliciously Easy Indian Food at Home*, Viva Books, Delhi, 2008.

Manoj Negi: *Career in Dairy Farming*, Abhishek Publication, Delhi, 2010.

P R Gupta:*Dairy India 2007: Serving the Dairy Industry Since 1983*, Dairy India Yearbook, 2007.

P T Rajan : *A Field Guide to Marine Food Fishes of Andaman and Nicobar Islands*, Zoological Survey of India, Delhi, 2003.

P.C. Trivedi:*Microbes : Applications and Effects*, Aavishkar Publication, Delhi, 2009.

Piyush Sharma and Prem Ram : *A Textbook of Food and Beverage Management*, Dominant Publication, Delhi, 2013.

R S Paroda: *Sustaining Our Food Security*, Konark Publication, Delhi, 2003.

R. P. Saxena.: *Hotel Management : Food And Food Services*, Delhi, Centrum Press, 2010.

S.K. Singh: *Fundamental of Hotel Management and Operations*, Delhi, Centrum Press, 2010.

S.K. Singh: *Strategic Management in Hotel and Restaurant Industry*, Delhi, Centrum Press, 2010.

Tapeshwar Singh: *Resource Conservation and Food Security : An Indian Experience (2 Vols-Set)*, Concept Publication, New Delhi, 2004.

Tharakan : *A Young Hotelier's Guide to Food and Beverage*, Tata McGraw-Hill Publication, Delhi, 2005.

Udai Veer: *Elements of Food Science*, Anmol Publications, Delhi, 2007.

Urvashi Nandal : *A Handbook of Foods and Nutritional Biochemistry*, Agrobios Publication, Delhi, 2013.

Vishwambhar Prasad Sati: *Natural Resources Conservation and Food Security*, Bishen Singh Mahendra Pal Singh, Dehra Dun, 2012.

W. Merback and A. Veha : *Crop Science and Technology for Food Security Bioenergy and Sustainability*, L. Bona, J. Pauk, ,

W.L. Slatter, T. Krishtofferson and D.A. Seiberling: *Manual of Dairy Industry : Training and Development of Personnel for the Dairy Industry*, Asiatic Publication, Delhi, 2012.

Index